INVESTIGATIONS IN NUMBER, DATA, AN[D SPACE]

Fractions

Fair Shares

Grade 3

Also appropriate for Grade 4

Cornelia C. Tierney
Mary Berle-Carman

Developed at TERC, Cambridge, Massachusetts

Dale Seymour Publications

The *Investigations* curriculum was developed at TERC (formerly Technical Education Research Centers) in collaboration with Kent State University and the State University of New York at Buffalo. The work was supported in part by National Science Foundation Grant No. MDR-9050210. TERC is a nonprofit company working to improve mathematics and science education. TERC is located at 2067 Massachusetts Avenue, Cambridge, MA 02140.

This project was supported, in part,
by the
National Science Foundation
Opinions expressed are those of the authors
and not necessarily those of the Foundation

This book is published by Dale Seymour Publications, an imprint of the Alternative Publishing Group of Addison-Wesley Publishing Company.

Project Editor: Priscilla Cox Samii
Series Editor: Beverly Cory
Manuscript Editor: Adrienne Harris
ESL Consultant: Nancy Sokol Green
Production/Manufacturing Director: Janet Yearian
Production/Manufacturing Coordinator: Barbara Atmore
Design Manager: Jeff Kelly
Design: Don Taka
Illustrations: Jane McCreary, Carl Yoshihara
Cover: Bay Graphics
Composition: Publishing Support Services

Printed on Recycled Paper

DALE SEYMOUR PUBLICATIONS
P.O. BOX 10888
PALO ALTO, CA 94303

Order number DS21245
ISBN 0-86651-807-X
1 2 3 4 5 6 7 8 9 10-ML-98 97 96 95 94

TERC

INVESTIGATIONS IN NUMBER, DATA, AND SPACE

Principal Investigator **Susan Jo Russell**

Co-Principal Investigator **Cornelia C. Tierney**

Director of Research and Evaluation **Jan Mokros**

Curriculum Development

Joan Akers
Michael T. Battista
Mary Berle-Carman
Douglas H. Clements
Karen Economopoulos
Ricardo Nemirovsky
Andee Rubin
Susan Jo Russell
Cornelia C. Tierney
Amy Shulman Weinberg

Evaluation and Assessment

Mary Berle-Carman
Abouali Farmanfarmaian
Jan Mokros
Mark Ogonowski
Amy Shulman Weinberg
Tracey Wright
Lisa Yaffee

Teacher Development and Support

Rebecca B. Corwin
Karen Economopoulos
Tracey Wright
Lisa Yaffee

Technology Development

Michael T. Battista
Douglas H. Clements
Julie Sarama Meredith
Andee Rubin

Video Production

David A. Smith

Administration and Production

Amy Catlin
Amy Taber

Cooperating Classrooms for This Unit

Katie Bloomfield
Robert A. Dihlmann
Shutesbury Elementary, Shutesbury, MA

Joan Forsyth
Jeanne Wall
Arlington Public Schools, Arlington, MA

Corrine Varon
Cambridge Public Schools, Cambridge, MA

Angela Philactos
Marjorie Tkacik
Meg Watson
Boston Public Schools, Boston, MA

Consultants and Advisors

Elizabeth Badger
Deborah Lowenberg Ball
Marilyn Burns
Ann Grady
Joanne M. Gurry
James J. Kaput
Steven Leinwand
Mary M. Lindquist
David S. Moore
John Olive
Leslie P. Steffe
Peter Sullivan
Grayson Wheatley
Virginia Woolley
Anne Zarinnia

Graduate Assistants

Kent State University:
Joanne Caniglia, Pam DeLong, Carol King

State University of New York at Buffalo:
Rosa Gonzalez, Sue McMillen,
Julie Sarama Meredith, Sudha Swaminathan

Investigations in Number, Data, and Space is a K–5 mathematics curriculum with four major goals:

- to offer students meaningful mathematical problems
- to emphasize depth in mathematical thinking rather than superficial exposure to a series of fragmented topics
- to communicate mathematics content and pedagogy to teachers
- to substantially expand the pool of mathematically literate students

The *Investigations* curriculum embodies an approach radically different from the traditional textbook-based curriculum. At each grade level, it consists of a set of separate units, each offering 2–4 weeks of work. These units of study are presented through investigations that involve students in the exploration of major mathematical ideas.

Approaching the mathematics content through investigations helps students develop flexibility and confidence in approaching problems, fluency in using mathematical skills and tools to solve problems, and proficiency in evaluating their solutions. Students also build a repertoire of ways to communicate about their mathematical thinking, while their enjoyment and appreciation of mathematics grows.

The investigations are carefully designed to invite all students into mathematics—girls and boys, diverse cultural, ethnic, and language groups, and students with different strengths and interests. Problem contexts often call on students to share experiences from their family, culture, or community. The curriculum eliminates barriers—such as work in isolation from peers, or emphasis on speed and memorization—that exclude some students from participating successfully in mathematics. The following aspects of the curriculum ensure that all students are included in significant mathematics learning:

- Students spend time exploring problems in depth.
- They find more than one solution to many of the problems they work on.
- They invent their own strategies and approaches, rather than relying on memorized procedures.
- They choose from a variety of concrete materials and appropriate technology, including calculators, as a natural part of their everyday mathematical work.
- They express their mathematical thinking through drawing, writing, and talking.
- They work in a variety of groupings—as a whole class, individually, in pairs, and in small groups.
- They move around the classroom as they explore the mathematics in their environment and talk with their peers.

While reading and other language activities are typically given a great deal of time and emphasis in elementary classrooms, mathematics often does not get the time it needs. If students are to experience mathematics in depth, they must have enough time to become engaged in real mathematical problems. We believe that a minimum of five hours of mathematics classroom time a week—about an hour a day—is critical at the elementary level. The plan and pacing of the Investigations curriculum is based on that belief.

For further information about the pedagogy and principles that underlie these investigations, see the overview book, *The Investigations Curriculum: Bringing Together Students, Teachers, and Mathematics*, as well as Teacher Notes throughout the units.

The *Investigations* curriculum is presented through a series of teacher books, one for each unit of study. These books not only provide a complete mathematics curriculum for your students, they offer materials to support your own professional development. You, the teacher, are the person who will make this curriculum come alive in the classroom; the book for each unit is your main support system.

While reproducible resources for students are provided, the curriculum does not include student books. Students work actively with objects and experiences in their own environment and with a variety of manipulative materials and technology, rather than with workbooks and worksheets filled with problems. We also make extensive use of the overhead projector as a way to present problems, to focus group discussion, and to help students share ideas and strategies. If an overhead projector is available, we urge you to try it as suggested in the investigations.

Ultimately, every teacher will use these investigations in ways that make sense for his or her particular style, the particular group of students, and the constraints and supports of a particular school environment. We have tried to provide with each unit the best information and guidance for a wide variety of situations, drawn from our collaborations with many teachers and students over many years. Our goal in this book is to help you, as a professional educator, implement this mathematics curriculum in a way that will give all your students access to mathematical power.

Investigation Format

The opening two pages of each investigation help you get ready for the student work that follows. Here you will read:

What Happens—a synopsis of each session or block of sessions.

Mathematical Emphasis—the most important ideas and processes students will encounter in this investigation.

What to Plan Ahead of Time—materials to gather, student sheets to duplicate, transparencies to make, and anything else you need to do before starting.

INVESTIGATION 1

Sharing Brownies

What Happens

Session 1 and 2: Making Fair Shares Students cut up rectangles as if they were brownies to share equally among a number of people. They show ways to share one brownie equally among 2 people, 4 people, 8 people, 3 people, and 6 people. They explore these same fractions by folding paper to make a set of Fraction Cards. They start to develop a class list of fraction facts.

Sessions 3 and 4: More Brownies to Share Students develop their own strategies to share more than one brownie equally among a number of people. They cut and paste paper brownies to show each person's share, and discuss different ways of naming shares of different sizes. They also add to the class list of fraction facts.

Mathematical Emphasis

- Realizing that fractional parts must be equal (for example, that one-third is not just one of three parts but one of three equal parts)
- Developing familiarity with conventional fraction words and notation (though students can write their solutions in any way that communicates accurately; for example a student might write $1/2 + 1/4$ as "half plus another piece that is half of the half")
- Becoming familiar with grouping unit fractions, those that have a numerator of 1 (for example, $1/6 + 1/6 + 1/6$ is equivalent to $3/6$, and $1/4 + 1/4 = 2/4$)

14 ■ *Investigation 1: Sharing Brownies*

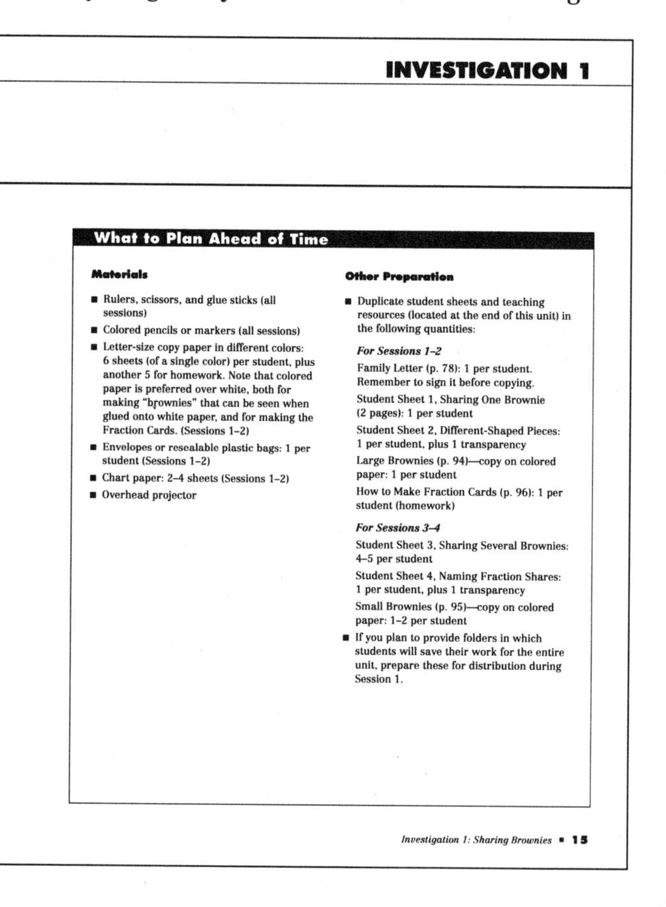

INVESTIGATION 1

What to Plan Ahead of Time

Materials

- Rulers, scissors, and glue sticks (all sessions)
- Colored pencils or markers (all sessions)
- Letter-size copy paper in different colors: 6 sheets (of a single color) per student, plus another 5 for homework. Note that colored paper is preferred over white, both for making "brownies" that can be seen when glued onto white paper, and for making the Fraction Cards. (Sessions 1–2)
- Envelopes or resealable plastic bags: 1 per student (Sessions 1–2)
- Chart paper: 2–4 sheets (Sessions 1–2)
- Overhead projector

Other Preparation

- Duplicate student sheets and teaching resources (located at the end of this unit) in the following quantities:

  For Sessions 1–2

  Family Letter (p. 78): 1 per student. Remember to sign it before copying.

  Student Sheet 1, Sharing One Brownie (2 pages): 1 per student

  Student Sheet 2, Different-Shaped Pieces: 1 per student, plus 1 transparency

  Large Brownies (p. 94)—copy on colored paper: 1 per student

  How to Make Fraction Cards (p. 96): 1 per student (homework)

  For Sessions 3–4

  Student Sheet 3, Sharing Several Brownies: 4–5 per student

  Student Sheet 4, Naming Fraction Shares: 1 per student, plus 1 transparency

  Small Brownies (p. 95)—copy on colored paper: 1–2 per student
- If you plan to provide folders in which students will save their work for the entire unit, prepare these for distribution during Session 1.

Investigation 1: Sharing Brownies ■ 15

Sessions Within an investigation, the activities are organized by class session, a session being a one-hour math class. Sessions are numbered consecutively through an investigation. Often several sessions are grouped together, presenting a block of activities with a single major focus.

When you find a block of sessions presented together—for example, Sessions 1, 2, and 3—read through the entire block first to understand the overall flow and sequence of the activities. Make some preliminary decisions about how you will divide the activities into three sessions for your class, based on what you know about your students. You may need to modify your initial plans as you progress through the activities, and you may want to make notes in the margins of the pages as reminders for the next time you use the unit.

Be sure to read the Session Follow-Up section at the end of the session block to see what homework assignments and extensions are suggested as you make your initial plans.

While you may be used to a curriculum that tells you exactly what each class session should cover, we have found that the teacher is in a better position to make these decisions. Each unit is flexible and may be handled somewhat differently by every teacher. While we provide guidance for how many sessions a particular group of activities is likely to need, we want you to be active in determining an appropriate pace and the best transition points for your class.

Ten-Minute Math At the beginning of some sessions, you will find Ten-Minute Math activities. These are designed to be used in tandem with the investigations, but not during the math hour. Rather, we hope you will do them whenever you have a spare 10 minutes—maybe before lunch or recess, or at the end of the day.

Ten-Minute Math offers practice in key concepts, but not always those being covered in the unit. For example, in a unit on using data, Ten-Minute Math might revisit geometric activities done earlier in the year. Complete directions for the suggested activities are included at the end of each unit. A compilation of Ten-Minute Math activities is also available as a separate book.

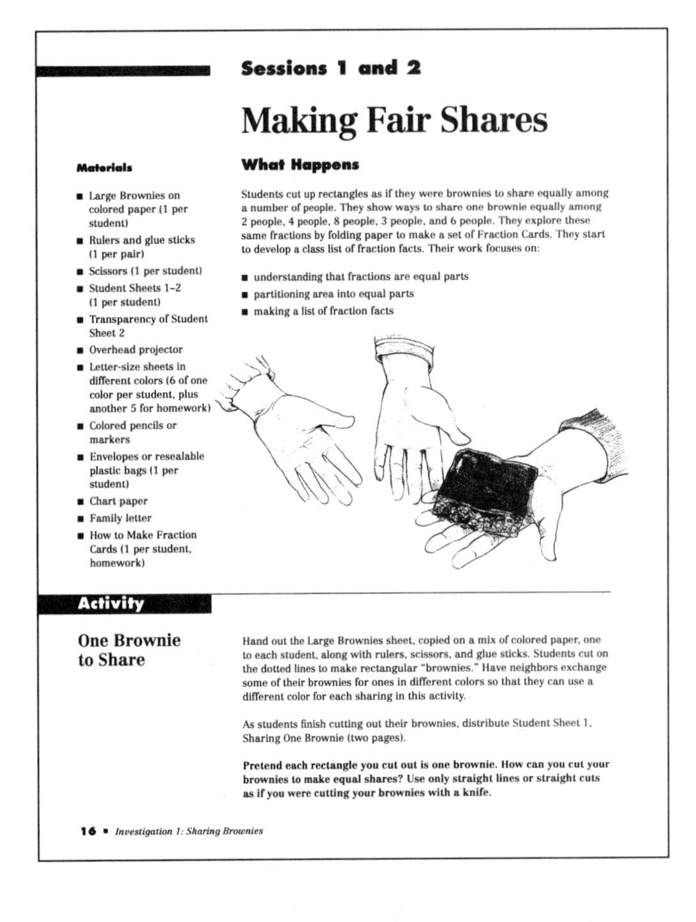

Sessions 1 and 2

Making Fair Shares

Materials

- Large Brownies on colored paper (1 per student)
- Rulers and glue sticks (1 per pair)
- Scissors (1 per student)
- Student Sheets 1–2 (1 per student)
- Transparency of Student Sheet 2
- Overhead projector
- Letter-size sheets in different colors (6 of one color per student, plus another 5 for homework)
- Colored pencils or markers
- Envelopes or resealable plastic bags (1 per student)
- Chart paper
- Family letter
- How to Make Fraction Cards (1 per student, homework)

What Happens

Students cut up rectangles as if they were brownies to share equally among a number of people. They show ways to share one brownie equally among 2 people, 4 people, 8 people, 3 people, and 6 people. They explore these same fractions by folding paper to make a set of Fraction Cards. They start to develop a class list of fraction facts. Their work focuses on:

- understanding that fractions are equal parts
- partitioning area into equal parts
- making a list of fraction facts

Activity

One Brownie to Share

Hand out the Large Brownies sheet, copied on a mix of colored paper, one to each student, along with rulers, scissors, and glue sticks. Students cut on the dotted lines to make rectangular "brownies." Have neighbors exchange some of their brownies for ones in different colors so that they can use a different color for each sharing in this activity.

As students finish cutting out their brownies, distribute Student Sheet 1, Sharing One Brownie (two pages).

Pretend each rectangle you cut out is one brownie. How can you cut your brownies to make equal shares? Use only straight lines or straight cuts as if you were cutting your brownies with a knife.

16 ▪ *Investigation 1: Sharing Brownies*

Activities The activities include pair and small-group work, individual tasks, and whole-class discussions. In any case, students are seated together, talking and sharing ideas during all work times. Students most often work cooperatively, although each student may record work individually.

Choice Time In some units, some sessions are structured with activity choices. In these cases, students may work simultaneously on different activities focused on the same mathematical ideas. Students choose which activities they want to do, and they cycle through them.

You will need to decide how to set up and introduce these activities and how to let students make their choices. Some teachers present them as station activities, in different parts of the room. Some list the choices on the board as reminders or have students keep their own lists.

Excursions One of the investigations in this unit includes an *excursion*—activities that could be omitted without harming the integrity of the unit. This is one way of dealing with the overabundance of fascinating mathematics to be studied—much more than a class has time to explore in any one year. Excursions give you the flexibility to make different choices from year to year. For example, you might do the excursion in this Fractions unit this year, but another year, try the excursions in another unit.

Tips for the Linguistically Diverse Classroom At strategic points in each unit, you will find concrete suggestions for simple modifications of the teaching strategies to encourage the participation of all students. Many of these tips offer alternative ways to elicit critical thinking from students at varying levels of English proficiency, as well as from other students who find it difficult to verbalize their thinking.

The tips are supported by suggestions for specific vocabulary work to help ensure that all students can participate fully in the investigations. The Preview for the Linguistically Diverse Classroom (p. 12) lists important words that are assumed as part of the working vocabulary of the unit. Second-language learners will need to become familiar with these words in order to understand the problems and activities they will be doing. These terms can be incorporated into students' second-language work before or during the unit. Activities that can be used to present the words are found in the appendix, Vocabulary Support for Second-Language Learners (p. 76).

In addition, ideas for making connections to students' language and cultures, included on the Preview page, help the class explore the unit's concepts from a multicultural perspective.

Session Follow-Up

Homework Homework is not given daily for its own sake, but periodically as it makes sense to have follow-up work at home. Homework may be used for (1) review and practice of work done in class (2) preparation for activities coming up, for example, collecting data for a class project; or (3) involving and informing family members.

Some units in the *Investigations* curriculum have more homework than others, simply because it makes sense for the mathematics that's going on. Other units rely on manipulatives that most students won't have at home, making homework difficult. In any case, homework should always be directly connected to the investigations in the unit, or to work in previous units—never sheets of problems just to keep students busy.

Extensions These follow-up activities are opportunities for some or all students to explore a topic in greater depth or in a different context. They are not designed for "fast" students; mathematics is a multifaceted discipline, and different students will want to go further in different investigations. Look for and encourage the sparks of interest and enthusiasm you see in your students, and use the extensions to help them pursue these interests.

Family Letter A letter that you can send home to students' families is included with the blackline masters for each unit. We want families to be informed about the mathematics work in your classroom; they should be encouraged to participate in and support their children's work. A reminder to send home the letter appears in one of the early investigations. (These letters are also available separately in the following languages: Spanish, Vietnamese, Cantonese, Hmong, and Cambodian.)

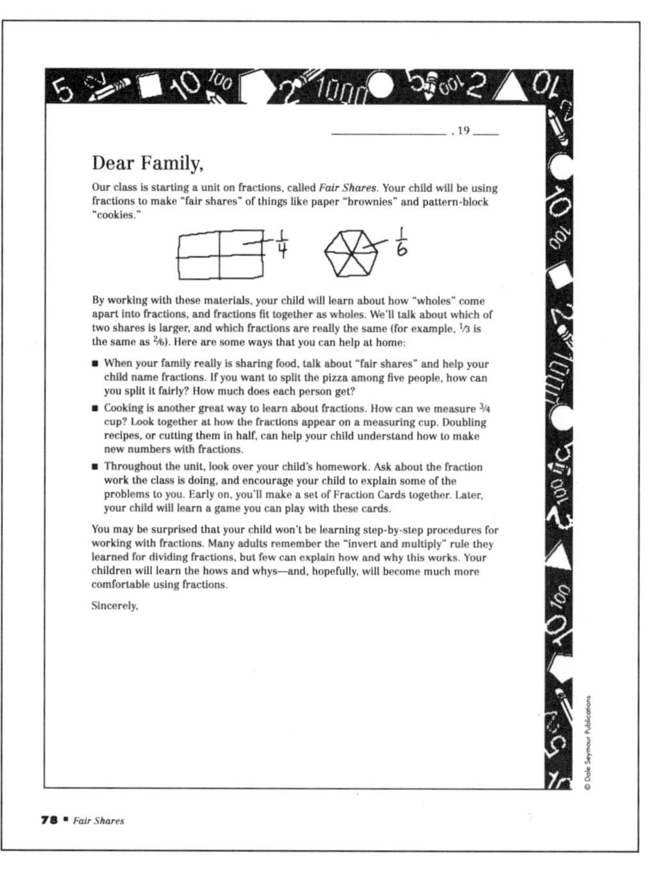

________________, 19____

Dear Family,

Our class is starting a unit on fractions, called *Fair Shares*. Your child will be using fractions to make "fair shares" of things like paper "brownies" and pattern-block "cookies."

By working with these materials, your child will learn about how "wholes" come apart into fractions, and fractions fit together as wholes. We'll talk about which of two shares is larger, and which fractions are really the same (for example, $\frac{1}{3}$ is the same as $\frac{2}{6}$). Here are some ways that you can help at home:

- When your family really is sharing food, talk about "fair shares" and help your child name fractions. If you want to split the pizza among five people, how can you split it fairly? How much does each person get?
- Cooking is another great way to learn about fractions. How can we measure $\frac{3}{4}$ cup? Look together at how the fractions appear on a measuring cup. Doubling recipes, or cutting them in half, can help your child understand how to make new numbers with fractions.
- Throughout the unit, look over your child's homework. Ask about the fraction work the class is doing, and encourage your child to explain some of the problems to you. Early on, you'll make a set of Fraction Cards together. Later, your child will learn a game you can play with these cards.

You may be surprised that your child won't be learning step-by-step procedures for working with fractions. Many adults remember the "invert and multiply" rule they learned for dividing fractions, but few can explain how and why this works. Your children will learn the hows and whys—and, hopefully, will become much more comfortable using fractions.

Sincerely,

78 ▪ *Fair Shares*

Materials

A complete list of the materials needed for the unit is found on p. 10. Some of these materials are available in a kit for the *Investigations* grade 3 curriculum. Individual items can also be purchased as needed from school supply stores and dealers.

In an active mathematics classroom, certain basic materials should be available at all times: interlocking cubes, pencils, unlined paper, graph paper, calculators, things to count with, and measuring tools. Some activities in this curriculum require scissors and glue sticks or tape. Stick-on notes and large paper are also useful materials throughout.

So that students can independently get what they need at any time, they should know where these materials are kept, how they are stored, and how they are to be returned to the storage area. For example, interlocking cubes are best stored in towers of ten; then, whatever the activity, they should be returned to storage in groups of ten at the end of the hour. You'll find that establishing such routines at the beginning of the year is well worth the time and effort.

Student Sheets and Teaching Resources

Reproducible pages to help you teach the unit are found at the end of this book. These include masters for making overhead transparencies and other teaching tools, as well as student recording sheets.

Many of the field-test teachers requested more sheets to help students record their work, and we have tried to be responsive to this need. At the same time, we think it's important that students find their own ways of organizing and recording their work. They need to learn how to explain their thinking with both drawings and written words, and how to organize their results so someone else can understand them.

To ensure that students get a chance to learn how to represent and organize their own work, we deliberately do not provide student sheets for every activity. We recommend that your students keep a mathematics notebook or folder so that their work, whether on reproducible sheets or their own paper, is always available to them for reference.

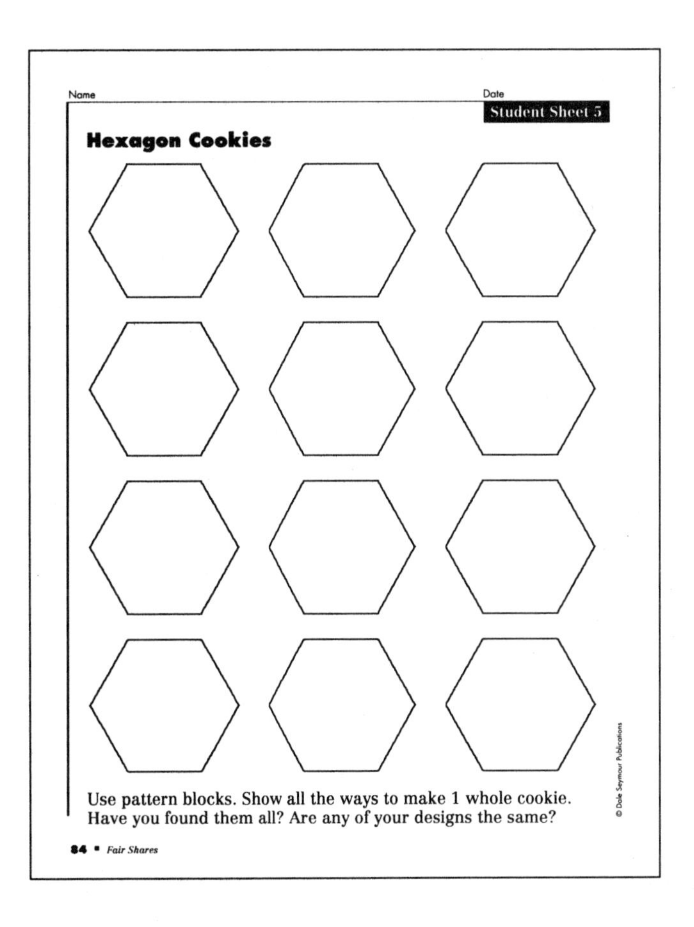

Name Date

Student Sheet 5

Hexagon Cookies

Use pattern blocks. Show all the ways to make 1 whole cookie.
Have you found them all? Are any of your designs the same?

84 ▪ *Fair Shares*

© Dale Seymour Publications

Help for You, the Teacher

Because we believe strongly that a new curriculum must help teachers think in new ways about mathematics and about their students' mathematical thinking processes, we have included a great deal of material to help you learn more about both.

About the Mathematics in This Unit This introductory section (p. 11) summarizes for you the critical information about the mathematics you will be teaching. This will be particularly valuable to teachers who are accustomed to a traditional textbook-based curriculum.

Teacher Notes These reference notes provide practical information about the mathematics you are teaching and about our experience with how students learn. Many of the notes were written in response to actual questions from teachers, or to discuss important things we saw happening in the field-test classrooms. Some teachers like to read them all before starting the unit, then review them as they come up in particular investigations.

Dialogue Boxes Sample dialogues throughout the unit demonstrate how students typically express their mathematical ideas, what issues and confusions arise in their thinking, and how some teachers have guided class discussions.

These dialogues are based on the extensive classroom testing of this curriculum; many are word-for-word transcriptions of recorded class discussions. They are not always easy reading; sometimes it may take some effort to unravel what the students are trying to say. But this is the value of these dialogues; they offer good clues to how your students may develop and express their approaches and strategies, helping you prepare for your own class discussions.

Where to Start You may not have time to read everything the first time you use this unit. As a first-time user, you will likely focus on understanding the activities and working them out with your students. Read completely through each investigation before starting to present it.

When you next teach this same unit, you can begin to read more of the background. Each time you present this unit, you will learn more about how your students understand the mathematical ideas. The first-time user of *Fair Shares* should read the following:

- About the Mathematics in This Unit (p. 11)
- Teacher Note: Different Shapes, Equal Pieces (p. 24)
- Dialogue Box: 7 Brownies, 4 People (p. 33)
- Teacher Note: Assessment: Sharing With and Without a Calculator (p. 66)

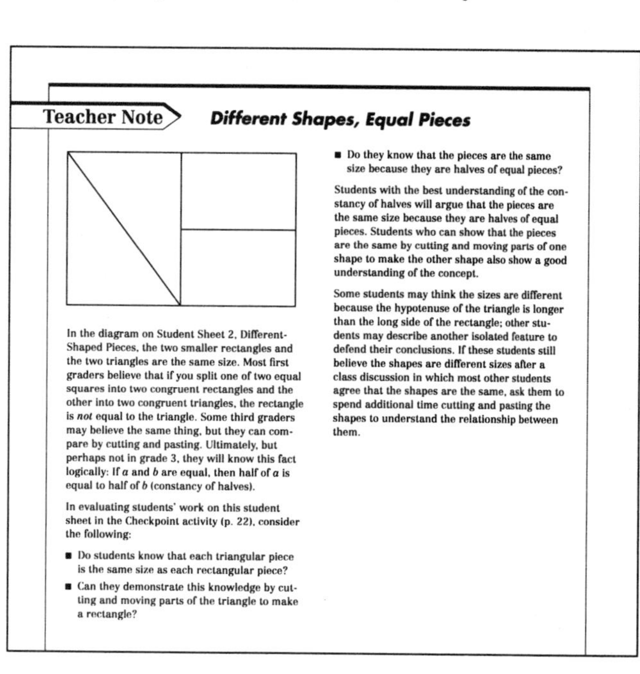

Teacher Note **Different Shapes, Equal Pieces**

In the diagram on Student Sheet 2, Different-Shaped Pieces, the two smaller rectangles and the two triangles are the same size. Most first graders believe that if you split one of two equal squares into two congruent rectangles and the other into two congruent triangles, the rectangle is *not* equal to the triangle. Some third graders may believe the same thing, but they can compare by cutting and pasting. Ultimately, but perhaps not in grade 3, they will know this fact logically: If a and b are equal, then half of a is equal to half of b (constancy of halves).

In evaluating students' work on this student sheet in the Checkpoint activity (p. 22), consider the following:

- Do students know that each triangular piece is the same size as each rectangular piece?
- Can they demonstrate this knowledge by cutting and moving parts of the triangle to make a rectangle?
- Do they know that the pieces are the same size because they are halves of equal pieces?

Students with the best understanding of the constancy of halves will argue that the pieces are the same size because they are halves of equal pieces. Students who can show that the pieces are the same by cutting and moving parts of one shape to make the other shape also show a good understanding of the concept.

Some students may think the sizes are different because the hypotenuse of the triangle is longer than the long side of the rectangle; other students may describe another isolated feature to defend their conclusions. If these students still believe the shapes are different sizes after a class discussion in which most other students agree that the shapes are the same, ask them to spend additional time cutting and pasting the shapes to understand the relationship between them.

DIALOGUE BOX

7 Brownies, 4 People

In this discussion of the Sharing Seven Brownies activity (p. 26), students talk their way through the dilemma of how to divide seven brownies equally among four people.

Amanda: First each person would get one.

Chantelle: But because there were only three left, and seven is an odd number—

Amanda: We give half to each person.

Chantelle: And have one left, and you can't divide it in half because there are four people.

Amanda: So we divide it into quarters and add a quarter to each half. That is one and a third.

This is what one-third looks like [*draws a whole rectangle and splits it into thirds*], and one and one-third would look like a whole plus one of these pieces.

Amanda: Oh. So what would half plus a fourth be?

That is the question! Let's draw what each person would get. Draw me one-half plus one-fourth.

Amanda [*Drawing a new picture in which she shades one-half and one-fourth in the same whole*]: Oh, it is three-fourths.

Chantelle: That makes sense, Amanda, because there are two-fourths in one-half, remember from yesterday?

Amanda: So each person gets one and a half and a quarter.

Jeremy: Give one whole to each person, then split the three brownies into four parts, one for each person.

Tell me what I should do to the picture.

Jeremy: Split the three brownies with a line this way [*makes vertical motion*], and this way [*makes horizontal motion*]. Now each person gets a piece from each brownie.

What size pieces are these?

Students: Quarters. Fourths.

So, how much brownie does each person get?

Jeremy: One whole brownie and three of the fourths pieces.

[*Teacher writes the following on the board:*]

one whole and three of the fourths

$1\frac{3}{4}$

Does anyone have another way?

Elena: You divide each of the brownies into four parts.

Yoshi: Yeah, that's how I was thinking about it. You give seven pieces to each person.

Elena: Each person gets seven of the fourths.

[*Teacher writes on the board:*]

$$\frac{1}{4} + \frac{1}{4} + \frac{1}{4} + \frac{1}{4} + \frac{1}{4} + \frac{1}{4} + \frac{1}{4} = \frac{7}{4}$$

Teacher Checkpoints As a teacher of the *Investigations* curriculum, you observe students daily, listen to their discussions, look carefully at their work, and use this information to guide your teaching. We have designated Teacher Checkpoints as natural times to get an overall sense of how your class is doing in the unit.

The Teacher Checkpoints provide a time for you to pause and reflect on your teaching plan while observing students at work in an activity. These sections offer tips on what you should be looking for and how you might adjust your pacing. Are most students fluent with strategies for solving a particular kind of problem? Are they just starting to formulate good strategies? Or are they still struggling with how to start?

Depending on what you see as the students work, you may want to spend more time on similar problems, change some of the problems to use smaller numbers, move quickly to more challenging material, modify subsequent activities for some students, work on particular ideas with a small group, or pair students who have good strategies with those who are having more difficulty.

In *Fair Shares* you will find two Teacher Checkpoints:

Different-Shaped Pieces (p. 22)
Writing About Shares (p. 29)

Embedded Assessment Activities Use the built-in assessments included in this unit to help you examine the work of individual students, figure out what it means, and provide feedback. From the students' point of view, the activities you will be using for assessment are no different from any others; they don't look or feel like traditional tests.

These activities sometimes involve writing and reflecting, at other times a brief interaction between student and teacher, and in still other instances the creation and explanation of a product.

In *Fair Shares* you will find assessment activities in the second and third investigations:

Letter to a Second Grader (p. 44)
Sharing With and Without a Calculator (p. 65)

Teachers find the hardest part of the assessment to be interpreting their students' work. If you have used a process approach to teaching writing, you will find our mathematics approach familiar. To help with interpretation, we provide guidelines and questions to ask about the students' work. In some cases we include a Teacher Note with specific examples of student work and a commentary on what it indicates. This framework can help you determine how your students are progressing.

As you evaluate students' work, it's important to remember that you're looking for much more than the "right answer." You'll want to know what their strategies are for solving the problem, how well these strategies work, whether they can keep track of and logically organize an approach to the problem, and how they make use of representations and tools to solve the problem.

Ongoing Assessment Good assessment of student work involves a combination of approaches. Some of the things you might do on an ongoing basis include the following:

- **Observation** Circulate around the room to observe students as they work. Watch for the development of their mathematical strategies, and listen to their discussions of mathematical ideas.
- **Portfolios** Ask students to document their work, in journals, notebooks, or portfolios. Periodically review this work to see how their mathematical thinking and writing are changing. Some teachers have students keep a notebook or folder for each unit, while others prefer one mathematics notebook or a portfolio of selected work for the entire year. Take time at the end of each unit for students to choose work for their portfolios. You might also have them write about what they've learned in the unit.

Fair Shares

OVERVIEW

Content of This Unit Students use fractions and mixed numbers as they solve sharing problems and build wholes from fractional parts. They find ways to share rectangular "brownies" and hexagonal pattern-block "cookies" among different numbers of people, and they play games that call for them to share these items. Students decide which of two shares is larger—for example, if three brownies are shared between two people or four brownies are shared among three people. Finally, students connect the sharing problems to division. They decide how to share things (such as balloons) that we cannot cut up into exact shares, and how to share dollars by converting to cents. They explore how to do sharing problems with a calculator, thus seeing their fraction answers as decimals. Throughout the unit, students develop a class chart of "fractions facts" like these:

$\frac{3}{4} = \frac{1}{2} + \frac{1}{4}$ $\frac{1}{6} + \frac{1}{3} = \frac{1}{2}$

In the process, they become familiar with reading conventional fraction notation.

Connections with Other Units If you are doing the full-year *Investigations* curriculum in the suggested sequence for grade 3, this is the ninth of ten units. In this unit, students build on their understanding of division, which they began to develop in the Multiplication and Division unit, *Things That Come in Groups. Fair Shares* can be used successfully at either grade 3 or grade 4, depending on the previous experience and needs of your students. The grade 4 Fractions and Area unit, *Different Shapes, Equal Pieces*, continues and extends the work begun in *Fair Shares*.

Investigations Curriculum ■ Suggested Grade 3 Sequence

Mathematical Thinking at Grade 3 (Introduction)

Things That Come in Groups (Multiplication and Division)

Flips, Turns, and Area (2-D Geometry)

From Paces to Feet (Measuring and Data)

Landmarks in the Hundreds (The Number System)

Up and Down the Number Line (Changes)

Combining and Comparing (Addition and Subtraction)

Turtle Paths (2-D Geometry)

▶ *Fair Shares* (Fractions)

Exploring Solids and Boxes (3-D Geometry)

Investigation 1 • Sharing Brownies

Class Sessions	Activities	Pacing	Ten-Minute Math
Sessions 1 and 2 MAKING FAIR SHARES	One Brownie to Share Making Fraction Cards From Smallest to Largest Finding Fraction Facts ■ Teacher Checkpoint: Different-Shaped Pieces ■ Homework ■ Extension	2 hrs	Guess My Number
Sessions 3 and 4 MORE BROWNIES TO SHARE	Sharing Seven Brownies More Sharing Problems ■ Teacher Checkpoint: Writing About Shares Comparing Names of Shares ■ Homework ■ Extension	2 hrs	

Investigation 2 • Pattern-Block Cookies

Class Sessions	Activities	Pacing	Ten-Minute Math
Sessions 1 and 2 MAKING COOKIES SHARES	Cutting Up Cookies What Fractions Can You Give Away Making Shares in Many Ways More Fraction Facts ■ Homework	2 hrs	Broken Calculator
Session 3 COMPARING SHARES	Reviewing Homework Which Is More? ■ Assessment: Letter to a Second Grader	1 hr	
Session 4 THE FRACTION COOKIE GAME	Game Warm-Up Playing the Fraction Cookie Game The Fraction Card Game ■ Homework	1 hr	
Sessions 5 and 6 BACKWARD SHARING	How Many Cookies in All? How Many People? How Many Cookies?	2 hrs	
Session 7 (Excursion)* HALF YELLOW	Designs That Are Half Yellow Other Fractions of Yellow How Do You Know It Is Half Yellow? ■ Extension	1 hr	

* Excursions can be omitted without harming the integrity or continuity of the unit, but offer good mathematical work if you have time to include them.

Continued on next page

Investigation 3 • Other Things to Share			
Class Sessions	**Activities**	**Pacing**	**Ten-Minute Math**
Sessions 1 and 2 HOW CAN WE SPLIT BALLOONS?	Things That Can't Be Shared Exactly Can We Split It? Sharing Dollars Fractions and Decimals ■ Assessment: Sharing With and Without a Calculator ■ Homework	2 hrs	Broken Calculator
Session 3 SHARING MANY THINGS	How Many for Each Person? ■ Extension	1 hr	

MATERIALS LIST

Following are the basic materials needed for the activities in this unit. The suggested quantities are ideal; however, in some instances you can work with smaller quantities by running several activities, requiring different materials, simultaneously.

Items marked with an asterisk are provided with the *Investigations* Materials Kit for grade 3.

* Pattern blocks: 1 bucket per 4–5 students

* Fraction dice in two colors: 3 per pair of students, 2 of one color, 1 of the other (blank dice or wooden inch cubes, marked with fractions, can be substituted)

* Rulers: 1 per pair

* Play money—dollars, quarters, and dimes (optional)

Items you could give to students to share, such as peanuts in shells, pennies, or paper clips: about 8–10 per student

Letter-size sheets in five colors: 10–12 per student

Colored pencils, markers, or crayons: yellow, red, blue, and green

Glue sticks: 1 per pair

Scissors: 1 per student

Calculators: 1 per student

Envelopes or resealable plastic bags: 1 per student

Chart paper: 2–4 sheets

Stick-on notes: 1–2 pads, small

Overhead projector

The following materials are provided at the end of this unit as blackline masters. They are also available as classroom sets.

Student Sheets 1–12

Teaching Resources:

Large Brownies

Small Brownies

How to Make Fraction Cards

Triangle Paper

Students need to have a sense of fractions and their relationships, just as they need number sense with whole numbers. Before learning to compute with fractions, they need to understand the fractions themselves. Traditional textbooks often ask students to label drawings that are already split into two or three or more equal parts. Because students do not learn to construct a half or a third or a sixth, they gain little sense of what these fractions are or the importance of equal parts. Students need to know that sixths are smaller than halves, for example, and have a mental image of how sixths can combine to make halves.

Understanding the relationship between sixths and halves requires more than finding common denominators; it calls for an important mathematical construction. Research has shown that many students in high school haven't developed meaning for fractions. As a result, they have no basis for checking their work or for choosing which of several memorized computation methods to use.

Students need familiarity with common equivalents, especially relationships between halves, fourths, and eighths, and between halves, thirds, and sixths. Fraction sense includes "seeing" automatically that two halves, three thirds, four fourths, and so on make one whole; and that two-fourths, three-sixths, or four-eighths of something are the same amount as one-half of the same thing—although a fraction of one whole is a different amount from the same fraction of a different-size whole. To see that describing a whole is a critical step in defining fractions of it, students need repeated experiences describing and comparing the sizes of things.

In this unit, students write about and illustrate their solutions to problems that require thinking about parts and wholes. They clarify their own reasoning and find images that help them visualize fractions. They split wholes into equal parts of their own making instead of identifying the fractional parts of an already split and shaded whole. Students become familiar with conventional fraction notation by connecting it to shape. They see how the teacher writes the fractional relationships they describe when they talk about parts of shapes.

Mathematical Emphasis At the beginning of each investigation, the Mathematical Emphasis section tells you what is most important for students to learn about during that investigation. Many of these mathematical understandings and processes are difficult and complex. Students gradually learn more and more about each idea over many years of schooling. Individual students will begin and end the unit with different levels of knowledge and skill, but all students will gain greater understanding of fractions and fraction notation.

In the *Investigations* curriculum, mathematical vocabulary is introduced naturally during the activities. We don't ask students to learn definitions of new terms; rather, they come to understand such words as *factor* or *area* or *symmetry* by hearing them used frequently in discussion as they investigate new concepts. This approach is compatible with current theories of second-language acquisition, which emphasize the use of new vocabulary in meaningful contexts while students are actively involved with objects, pictures, and physical movement.

Listed below are some key words used in this unit that will not be new to most English speakers at this age level, but may be unfamiliar to students with limited English proficiency. You will want to spend additional time working on these words with your students who are learning English. If your students are working with a second-language teacher, you might enlist your colleague's aid in familiarizing students with these words, before and during this unit. In the classroom, look for opportunities for students to hear and use these words. Activities you can use to present the words are given in the appendix, Vocabulary Support for Second-Language Learners (p. 76).

whole, piece, smallest, largest Throughout the unit, students consider and cut apart *wholes* (rectangles, hexagons, larger designs) into *pieces* of equal size. They compare the pieces of different wholes to determine which are the *largest* and *smallest*.

share, fair In sharing problems, students explore cutting up wholes into parts or *shares* that represent fractions of the whole. They see that to be *fair* shares, the parts must be equal in size.

brownie, cookie These terms are used to refer to simple rectangles and to small hexagon shapes (pattern blocks) to give meaning to the idea of "sharing" parts of them.

Multicultural Extensions for All Students

Whenever possible, encourage students to share words, objects, customs, or any aspects of daily life from their own cultures and backgrounds that are relevant to the activities in this unit. For example:

- When students are listing things that can and cannot be shared equally (Investigation 3), encourage them to include items that are shared in their families or cultures. Allow time for students to tell about shared items in their cultures that may be unfamiliar to others in the class. You might include celebrations that are "time-shared," perhaps by moving from home to home.

Investigations

INVESTIGATION 1

Sharing Brownies

What Happens

Session 1 and 2: Making Fair Shares Students cut up rectangles as if they were brownies to share equally among a number of people. They show ways to share one brownie equally among 2 people, 4 people, 8 people, 3 people, and 6 people. They explore these same fractions by folding paper to make a set of Fraction Cards. They start to develop a class list of fraction facts.

Sessions 3 and 4: More Brownies to Share Students develop their own strategies to share more than one brownie equally among a number of people. They cut and paste paper brownies to show each person's share, and discuss different ways of naming shares of different sizes. They also add to the class list of fraction facts.

Mathematical Emphasis

- Realizing that fractional parts must be equal (for example, that one-third is not just one of three parts but one of three equal parts)
- Developing familiarity with conventional fraction words and notation (though students can write their solutions in any way that communicates accurately; for example a student might write $1/2 + 1/4$ as "half plus another piece that is half of the half")
- Becoming familiar with grouping unit fractions, those that have a numerator of 1 (for example, $1/6 + 1/6 + 1/6$ is equivalent to $3/6$, and $1/4 + 1/4 = 2/4$)

What to Plan Ahead of Time

Materials

- Rulers, scissors, and glue sticks (all sessions)
- Colored pencils or markers (all sessions)
- Letter-size copy paper in different colors: 6 sheets (of a single color) per student, plus another 5 for homework. Note that colored paper is preferred over white, both for making "brownies" that can be seen when glued onto white paper, and for making the Fraction Cards. (Sessions 1–2)
- Envelopes or resealable plastic bags: 1 per student (Sessions 1–2)
- Chart paper: 2–4 sheets (Sessions 1–2)
- Overhead projector

Other Preparation

- Duplicate student sheets and teaching resources (located at the end of this unit) in the following quantities:

  For Sessions 1–2

  Family Letter (p. 78): 1 per student. Remember to sign it before copying.

  Student Sheet 1, Sharing One Brownie (2 pages): 1 per student

  Student Sheet 2, Different-Shaped Pieces: 1 per student, plus 1 transparency

  Large Brownies (p. 94)—copy on colored paper: 1 per student

  How to Make Fraction Cards (p. 96): 1 per student (homework)

  For Sessions 3–4

  Student Sheet 3, Sharing Several Brownies: 4–5 per student

  Student Sheet 4, Naming Fraction Shares: 1 per student, plus 1 transparency

  Small Brownies (p. 95)—copy on colored paper: 1–2 per student

- If you plan to provide folders in which students will save their work for the entire unit, prepare these for distribution during Session 1.

Sessions 1 and 2

Making Fair Shares

Materials

- Large Brownies on colored paper (1 per student)
- Rulers and glue sticks (1 per pair)
- Scissors (1 per student)
- Student Sheets 1–2 (1 per student)
- Transparency of Student Sheet 2
- Overhead projector
- Letter-size sheets in different colors (6 of one color per student, plus another 5 for homework)
- Colored pencils or markers
- Envelopes or resealable plastic bags (1 per student)
- Chart paper
- Family letter
- How to Make Fraction Cards (1 per student, homework)

What Happens

Students cut up rectangles as if they were brownies to share equally among a number of people. They show ways to share one brownie equally among 2 people, 4 people, 8 people, 3 people, and 6 people. They explore these same fractions by folding paper to make a set of Fraction Cards. They start to develop a class list of fraction facts. Their work focuses on:

- understanding that fractions are equal parts
- partitioning area into equal parts
- making a list of fraction facts

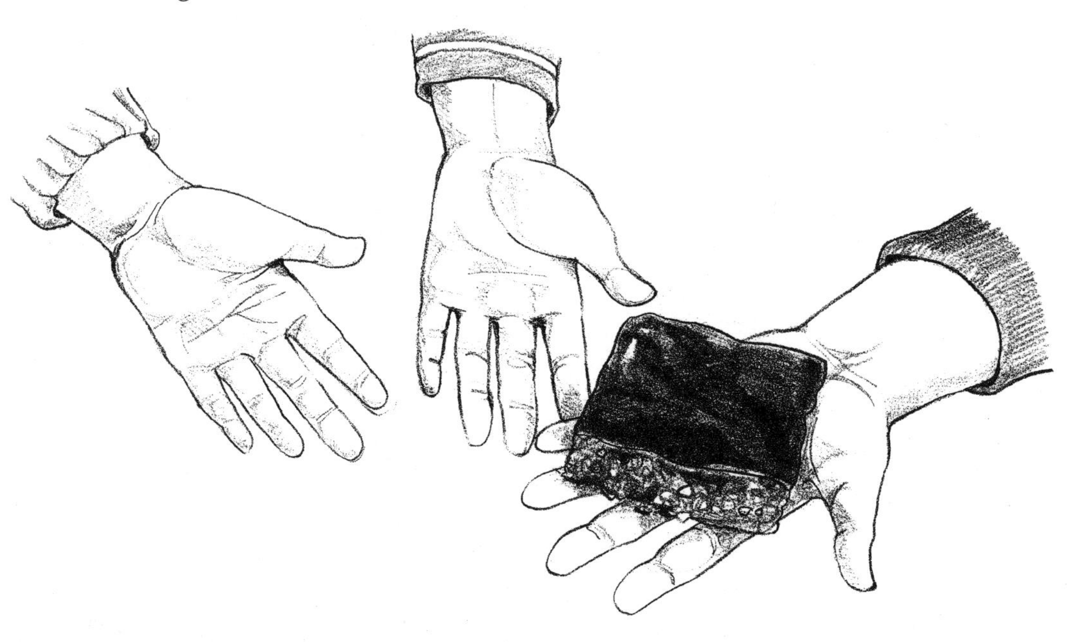

Activity

One Brownie to Share

Hand out the Large Brownies sheet, copied on a mix of colored paper, one to each student, along with rulers, scissors, and glue sticks. Students cut on the dotted lines to make rectangular "brownies." Have neighbors exchange some of their brownies for ones in different colors so that they can use a different color for each sharing in this activity.

As students finish cutting out their brownies, distribute Student Sheet 1, Sharing One Brownie (two pages).

Pretend each rectangle you cut out is one brownie. How can you cut your brownies to make equal shares? Use only straight lines or straight cuts as if you were cutting your brownies with a knife.

Students draw or cut three “brownies” (rectangles) into equal shares as called for on the first page of the student sheet—one into 2 shares, one into 4 shares, and one into 8 shares. Then they paste the pieces of their cut-apart brownies on the student sheet.

While the students are working, ask them to prove to you that all the shares of any given brownie are equal. Students may make equal shares that are rectangles or triangles. Partners might collaborate to try different ways of cutting up the brownies (for example, by cutting on the diagonal rather than on a vertical line).

Students exchange their work with another pair to check that they have equal halves, fourths, and eighths.

Now cut two more brownies into equal shares, again using straight lines. This time make 3 and 6 equal shares.

Students continue by cutting brownie-rectangles into thirds and sixths and pasting the pieces on the second page of the student sheet.

Collect the unused large brownie-rectangles, or let students save them for use in the Teacher Checkpoint activity at the end of this session.

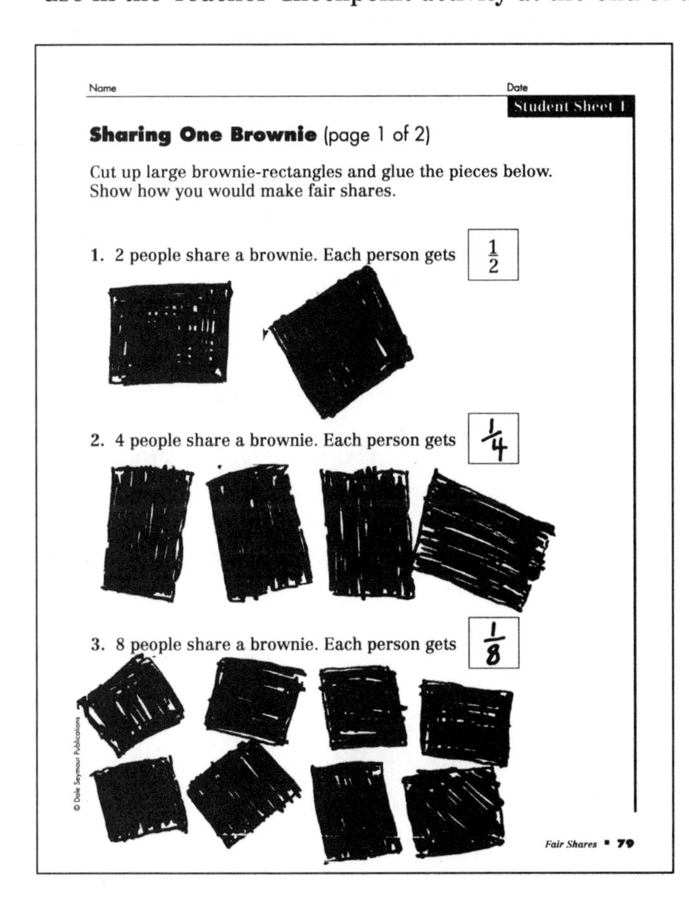
Name Date

Student Sheet 1

Sharing One Brownie (page 1 of 2)

Cut up large brownie-rectangles and glue the pieces below.
Show how you would make fair shares.

1. 2 people share a brownie. Each person gets $\frac{1}{2}$

2. 4 people share a brownie. Each person gets $\frac{1}{4}$

3. 8 people share a brownie. Each person gets $\frac{1}{8}$

Fair Shares ▪ 79

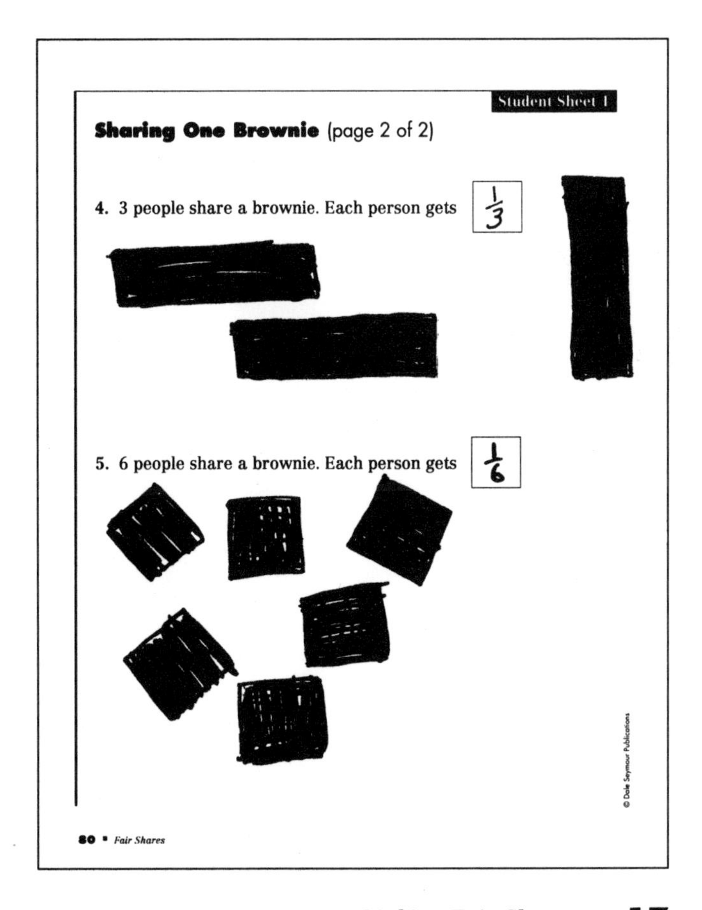
Student Sheet 1

Sharing One Brownie (page 2 of 2)

4. 3 people share a brownie. Each person gets $\frac{1}{3}$

5. 6 people share a brownie. Each person gets $\frac{1}{6}$

80 ▪ Fair Shares

Activity

Making Fraction Cards

Give each student five sheets of colored letter-size paper (all five sheets the same color). Explain that they will be using these sheets to make Fraction Cards. Then demonstrate the process. Fold a piece of paper in half, and mark the fold line with a marker. While you are doing this, ask students:

How many equal pieces am I making? Now you do the same with one of your sheets.

What fraction is each part?

Students fold one of their sheets in half, the same way you did, to show the two equal pieces. Demonstrate fraction notation for students by writing 1/2 in large numbers on each half of your model. (Leave the back of the sheet blank.) Post the sheet on the board with the labels showing. **Note:** To make it easier to recognize "top" and "bottom" numbers, write all fractions that students will see with the numerator directly above the denominator, and with the divide line horizontal rather than diagonal.

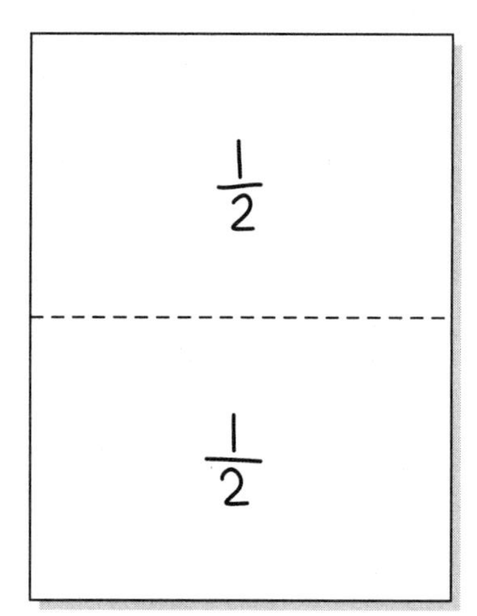

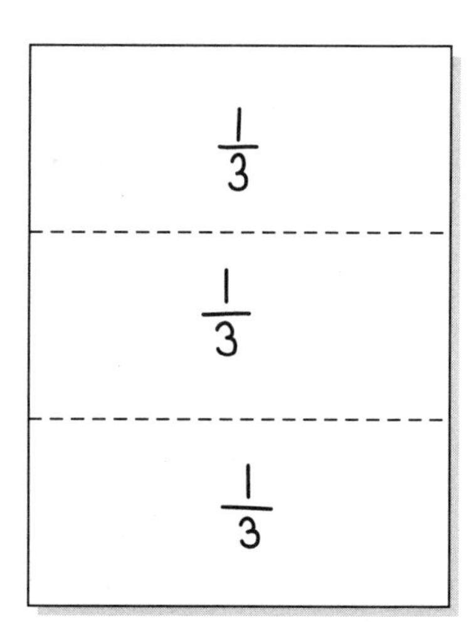

Fold a second sheet of paper into three equal parts, as if folding a letter to fit in an envelope. Mark the fold lines with a marker. Students also fold a sheet in this way. This task will be difficult for many students. Allow time for them to help one another. You might take your sheet around to measure with and mark on the edges of students' sheets where the folds should be. Students who can fold thirds accurately can fold pages for other students and make extras for students to use later in folding sixths.

When students each have a sheet folded in three parts, ask them what fraction each part is and how they think you should write it. When the class agrees that each piece is one-third, or one out of three, label the sections of your sheet with 1/3. Students do the same on their sheets.

Using the same process, but also making one vertical fold down the middle, make sheets of fourths and eighths, folding and labeling so that everyone can see. Each time, students fold their sheets as you do (using one sheet for fourths and one for eighths) and agree as a class about what fraction to write on each of the parts.

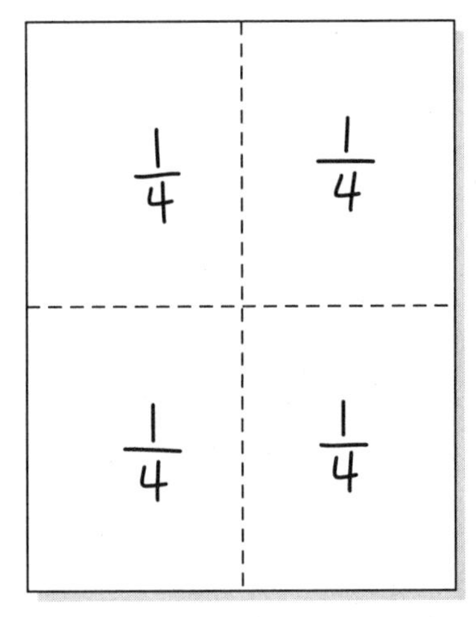

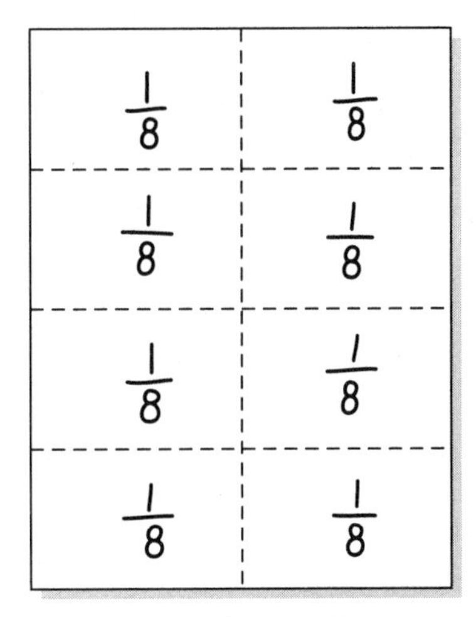

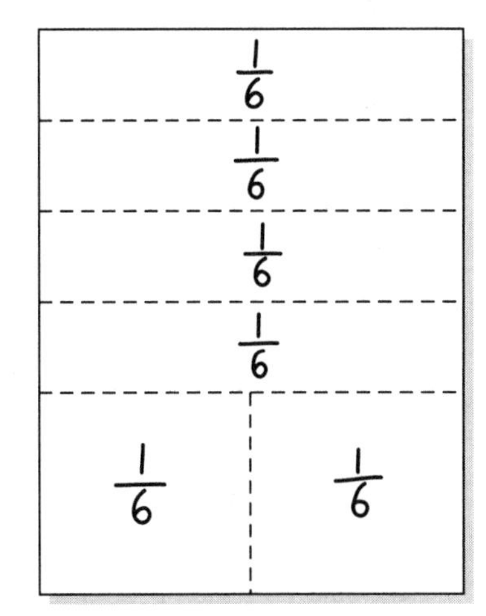

To make sixths, start by folding the sheet as you did for thirds. You may want to cut the thirds apart first, then make sixths by dividing them as follows: Fold *two* of the thirds in half with a lengthwise fold, making four long skinny strips. Fold the *other* third in half the short way, making two chunkier sixths. (The long strips make it easier to combine sixths with halves when making a whole.) Label each of the parts.

Students now have sheets folded and marked into halves, thirds, fourths, sixths, and eighths. They cut these apart on the marked lines to make their set of Fraction Cards. Distribute envelopes or resealable plastic bags to keep them in. Students will use the Fraction Cards in Ten-Minute Math. They will later make a second set at home for the Fraction Card game.

Skinny and Chunky Sixths Plan to spend some extra time discussing the different shapes they made for the sixths cards.

Are all six pieces, including the skinny ones and the chunky ones, the same size? How can we check to see if all the sixths are the same size?

Students work alone or in pairs. They may need to fold and cut their sixths to determine the answer. After a few minutes, some students may notice that if they cut the skinny sixth in half, the two pieces can be stacked to cover the chunky sixth exactly. Or they can cut the chunkier sixth in half and put the pieces end to end to cover the skinny sixth. One or two students can demonstrate their solutions to the class.

After cutting some of their sixths to prove that different-sized pieces have equal areas, students can tape them together again or make new sixths to be sure they end up with complete sets of Fraction Cards.

Activity

From Smallest to Largest

From the Fraction Cards you made in the preceding activity, select one of each size (use one of the chunky sixths). Spread your cards in random order, turned over so that students cannot see the fractions written on them. Place them within student reach—you might tape them on the board or tack them on a bulletin board.

I want to order these fractions from smallest to largest, with the smallest over here (to the left) and the largest over here (to the right).

Invite volunteers to rearrange the cards, moving one at a time, until the cards are arranged in order from smallest to largest.

Ask if anyone is not entirely sure the cards are in the correct order by size. Students who are uncertain may come up and check by placing one card on top of another to compare them directly. Sometimes the difference between sixths and eighths is difficult to see without a close inspection. When the class is sure that the cards are in order, ask students to try to name the fraction each card represents.

When students have guessed the names of the fractions, turn the cards over one by one to reveal the fraction. Students may notice a number pattern—each larger fraction has a smaller denominator than the one before—and use it to check that the cards are in order from smallest to largest.

Activity

Finding Fraction Facts

Begin a class list of fraction facts on chart paper. You might start two lists—one for the family of fraction facts using halves, fourths, and eighths, and another for the fraction facts using halves, thirds, and sixths.

What are some things you have noticed about fractions?

How can you combine two fractions to make a new fraction? How can you cut a fraction to make new fractions?

Here is a chance to introduce the conventional symbols for fractions with numerators larger than 1, as well as a chance to add fractions. Listen to your students' suggestions and write them using conventional notation. Present these notations not as the *only* correct way to write fractions, but as common usage. Be sure to write some of the students' suggestions in words; for example, "one-fourth is half of a half" or "four-eighths make a half."

Halves, Fourths, Eighths

$\frac{1}{4} + \frac{1}{4} = \frac{1}{2}$

8 eighths make a whole

Halves, Thirds, Sixths

$\frac{1}{3} + \frac{1}{3} + \frac{1}{3} = 1$

A sixth is half of a third

$\frac{2}{3} = \frac{4}{6}$

Third grade students are accustomed to using the equal sign only to point to an answer to a problem, not more generally to join two equal things. For example, they might be comfortable with $1/4 + 1/4 = 1/2$, but not with $2/4 = 1/2$. Give them time to talk about this difference. Introduce the equal sign as a way to join equal things. Encourage students to invent fraction sentences that join equal amounts in a variety of arrangements. For example:

$\frac{2}{4} = \frac{1}{2}$ $\qquad$ $\frac{1}{2} = \frac{1}{4} + \frac{1}{4}$ $\qquad$ $\frac{1}{2} + \frac{1}{4} + \frac{1}{4} = \frac{1}{2} + \frac{1}{2}$

$\frac{1}{4} = \frac{1}{2}$ of $\frac{1}{2}$ $\qquad$ $\frac{3}{3} = \frac{1}{3} + \frac{1}{3} + \frac{1}{3}$ $\qquad$ $\frac{6}{6} = 1 = \frac{3}{3} = \frac{4}{4} = \frac{8}{8}$

You will want to keep these lists posted throughout the unit. Students will add to the lists in future sessions as they discover new relationships among fractions. In Investigation 2, Pattern-Block Cookies, students will find more relationships among halves, thirds, and sixths.

Activity

Teacher Checkpoint

Different-Shaped Pieces

Use Student Sheet 2, Different-Shaped Pieces, to check whether students understand equality of area.

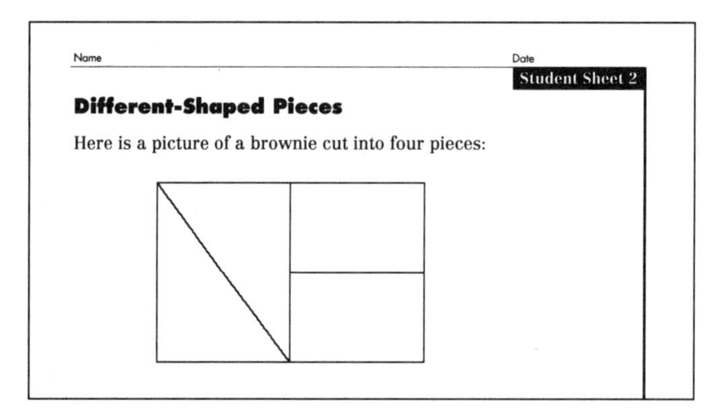
Name Date

Student Sheet 2

Different-Shaped Pieces

Here is a picture of a brownie cut into four pieces:

Looking at the above diagram of a paper brownie, students complete the statement "I believe that the rectangle pieces and the triangle pieces are/are not the same size because...." Students can cut and paste their own large paper brownies (using those left from earlier in the session) to see if the shapes are the same size. For help in evaluating students' work, see the **Teacher Note**, Different Shapes, Equal Pieces (p. 24).

After students have completed the checkpoint, put a transparency of Student Sheet 2 on the overhead or draw the picture of the brownie-rectangle on the board. Invite volunteers to show everyone how they know that the triangle is the same size as the rectangle. They might draw their ideas on the board or show the pieces of their cut-up brownie on the overhead projector.

If some students still don't believe the shapes are the same area, give them a right triangle (which you can make by folding and cutting a piece of paper on a diagonal) for them to cut and rearrange to make a rectangle.

Sessions 1 and 2 Follow-Up

Send home the family letter. Students also take home a copy of How to Make Fraction Cards and five sheets of colored paper (if possible), to make a second set of cards. Ideally, the second set should be a different color from the first set they made in class. Students can use this set of Fraction Cards at home throughout the unit. In Investigation 2, they will learn a fraction game they can play at home with their two sets of cards.

Extension

Ordering Larger Fractions Make another set of Fraction Cards, modified as follows:

Fold and mark each sheet as directed in Making Fraction Cards (p. 18), but before writing any labels, cut out one of the fractional parts of each sheet (cut out one of the chunky sixths). Label this part with its fraction ($1/2$, $1/3$, $1/4$, $1/6$, $1/8$), and label what remains of each large piece with its fraction ($1/2$, $2/3$, $3/4$, $5/6$, $7/8$).

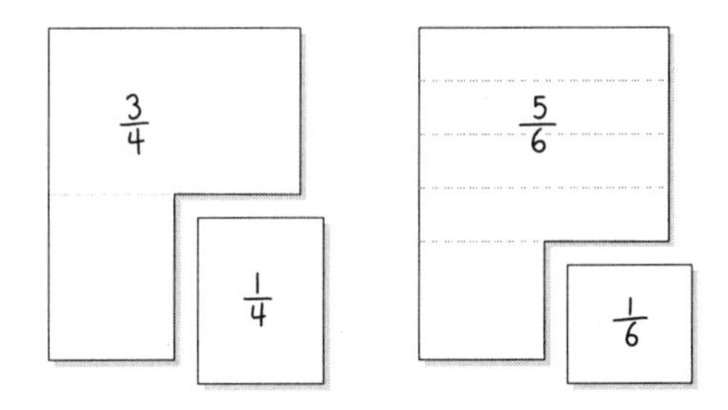

With student help, display all of the fraction pieces, both large and small, in order. Give students a minute to look for patterns among these ordered fractions; then cover the pieces or turn them so the labels don't show and ask students what they noticed. You might ask a few questions:

Which fractions are larger than $1/2$? Which are smaller than $1/2$?

Which is larger, $5/6$ or $3/4$? Each is missing one piece—why aren't they the same size?

Uncover the fraction labels and ask students to say more about what they notice about the ordered pieces.

Teacher Note

Different Shapes, Equal Pieces

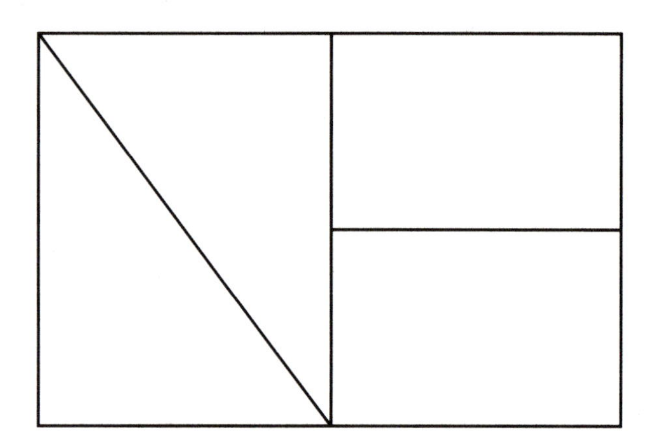

In the diagram on Student Sheet 2, Different-Shaped Pieces, the two smaller rectangles and the two triangles are the same size. Most first graders believe that if you split one of two equal squares into two congruent rectangles and the other into two congruent triangles, the rectangle is *not* equal to the triangle. Some third graders may believe the same thing, but they can compare by cutting and pasting. Ultimately, but perhaps not in grade 3, they will know this fact logically: If a and b are equal, then half of a is equal to half of b (constancy of halves).

In evaluating students' work on this student sheet in the Checkpoint activity (p. 22), consider the following:

- Do students know that each triangular piece is the same size as each rectangular piece?
- Can they demonstrate this knowledge by cutting and moving parts of the triangle to make a rectangle?
- Do they know that the pieces are the same size because they are halves of equal pieces?

Students with the best understanding of the constancy of halves will argue that the pieces are the same size because they are halves of equal pieces. Students who can show that the pieces are the same by cutting and moving parts of one shape to make the other shape also show a good understanding of the concept.

Some students may think the sizes are different because the hypotenuse of the triangle is longer than the long side of the rectangle; other students may describe another isolated feature to defend their conclusions. If these students still believe the shapes are different sizes after a class discussion in which most other students agree that the shapes are the same, ask them to spend additional time cutting and pasting the shapes to understand the relationship between them.

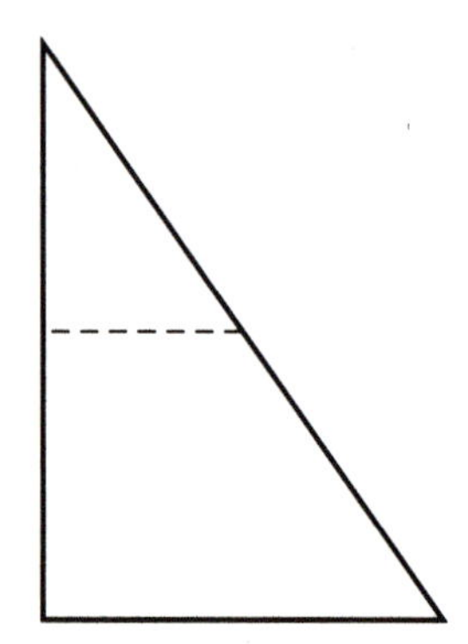

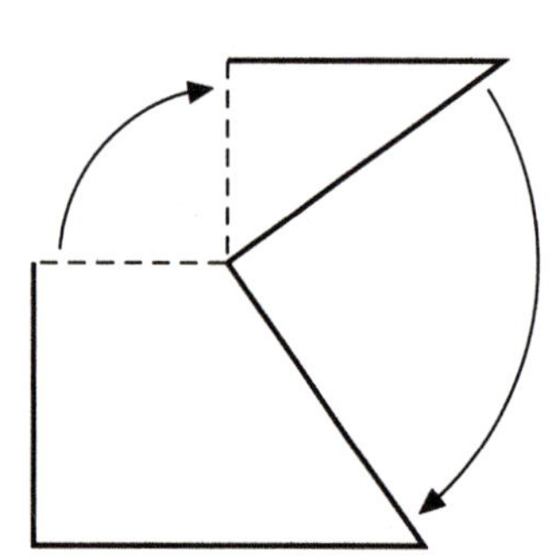

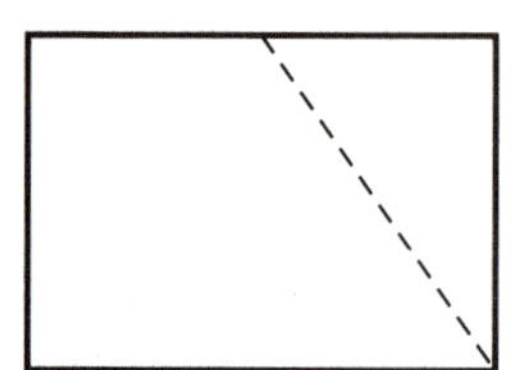

More Brownies to Share

What Happens

Students develop their own strategies to share more than one brownie equally among a number of people. They cut and paste paper brownies to show each person's share, and discuss different ways of naming shares of different sizes. They also add to the class lists of fraction facts. Their work focuses on:

- sharing several brownie-rectangles equally
- naming brownie-rectangle shares in fraction sentences

Materials

- Small Brownies on colored paper (1–2 per student)
- Student Sheet 3 (3–4 per student)
- Student Sheet 4 (1 per student)
- Transparency of Student Sheet 4
- Glue sticks and rulers (1 per pair)
- Scissors (1 per student)
- Fraction Cards (for Ten-Minute Math)

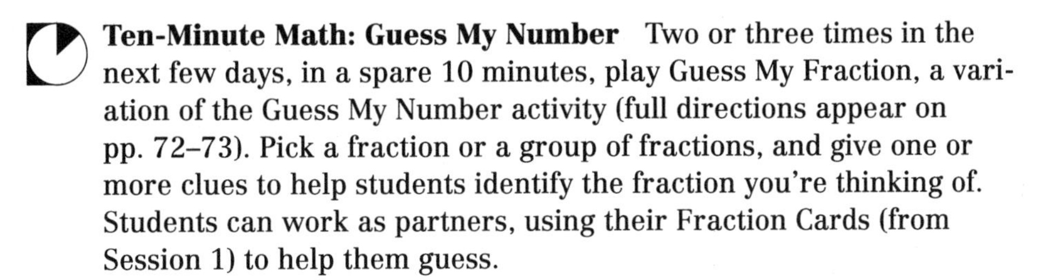

Ten-Minute Math: Guess My Number Two or three times in the next few days, in a spare 10 minutes, play Guess My Fraction, a variation of the Guess My Number activity (full directions appear on pp. 72–73). Pick a fraction or a group of fractions, and give one or more clues to help students identify the fraction you're thinking of. Students can work as partners, using their Fraction Cards (from Session 1) to help them guess.

Instead of calling out their answers, they write the fraction on a piece of paper and hold up the paper for you to see. Write all the students' ideas on the board. Then have students determine whether the answers fit the problem. Give another clue to narrow down the possible answer, or answer yes-or-no questions from students. When you demonstrate the game, ask students which fractions written on the board are no longer possibilities and cross them out.

For example, you might start by thinking about $\frac{1}{3}$ and giving clues as follows:

I'm thinking of a fraction. It is smaller than one-half. Talk with your partner to decide what fraction it could be.

Although there are infinite possibilities (such as $\frac{1}{5}$, $\frac{2}{5}$, $\frac{2}{7}$, $\frac{49}{100}$), students will most likely guess one of the fractions they have used: $\frac{1}{3}$, $\frac{1}{4}$, $\frac{1}{6}$, or $\frac{1}{8}$. After you write all the possibilities students suggest, give other clues one at a time:

It is larger than one-fourth.

Its numerator (top number) is 1.

With each new clue, ask students to eliminate the possibilities that are no longer in the running.

Activity

Sharing Seven Brownies

Introduce the following problem to the class at the board by drawing seven rectangles on a "plate" and four faces.

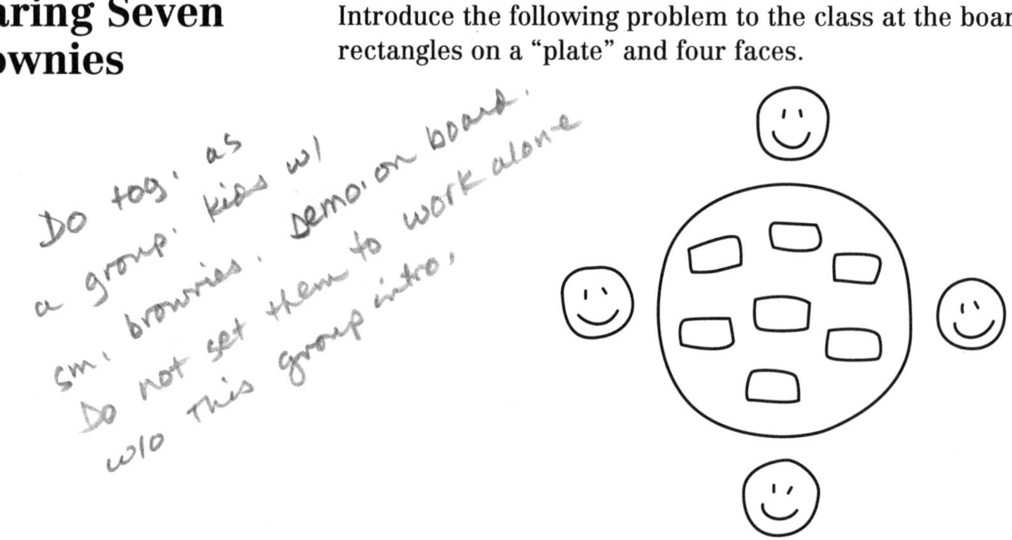

Imagine that you have seven brownies to share equally among four people. About how many brownies do you think each person will get? Do you think each person will get about one brownie? about two brownies? more than two brownies? See if you can find out exactly how many brownies each person will get. Be sure that each person gets exactly the same share.

Distribute the sheets of Small Brownies, duplicated on colored paper. Students need not cut apart the entire sheet at this time, but instead cut out the brownie-rectangles as needed for each activity in these sessions. Mention that these brownies are smaller than the ones students worked with in Session 1. Also hand out one copy of Student Sheet 3, Sharing Several Brownies.

Before starting, students fill in the blanks at the top of the sheet to represent the problem you have presented: 7 brownies shared by 4 people. They then cut and paste small brownies to show their solution on the student sheet, clearly showing each person's share.

Students can work alone or with partners. For this first problem, they cut up seven small brownie-rectangles to give equal amounts to each person. They group each person's share on the student sheet, and name (label) the share each person receives. In this example, possible names for one person's share are "$1 + 1/2 + 1/4$," or "seven-fourths ($7/4$)."

Some students may want to draw the number of people who share the brownies, or use a small paper plate to represent each person and deal out bits of brownies. Early finishers can check their work with others, or look for some fraction facts they could add to the posted lists.

Reporting Strategies for Sharing Ask for volunteers to describe how they solved the problem—how they started, and what they did next. Illustrate the different approaches on the board.

For example, if the student says, "First we cut all the brownies in half," divide your seven rectangles in half. Continue illustrating the student's method until you have divided all the brownies. Then ask students how they can be sure that each person received the same amount. After the class agrees that the solution is fair, ask how much each person receives. Write all the names that students suggest for each person's share, using fraction notation.

Collect on the board other strategies for approaching the problem, as well as other ways of recording the solution. See the **Dialogue Box**, 7 Brownies, 4 People (p. 33), for an excerpt from one class's discussion of their solutions.

While students are reporting their strategies, do not tell them whether their solutions are correct or incorrect. After the students have shown and explained their solutions, ask the rest of the class if they think the shares are fair. Students will correct mistakes themselves as they seek to explain their thinking and will incorporate each other's helpful comments. Students are likely to offer the following solutions:

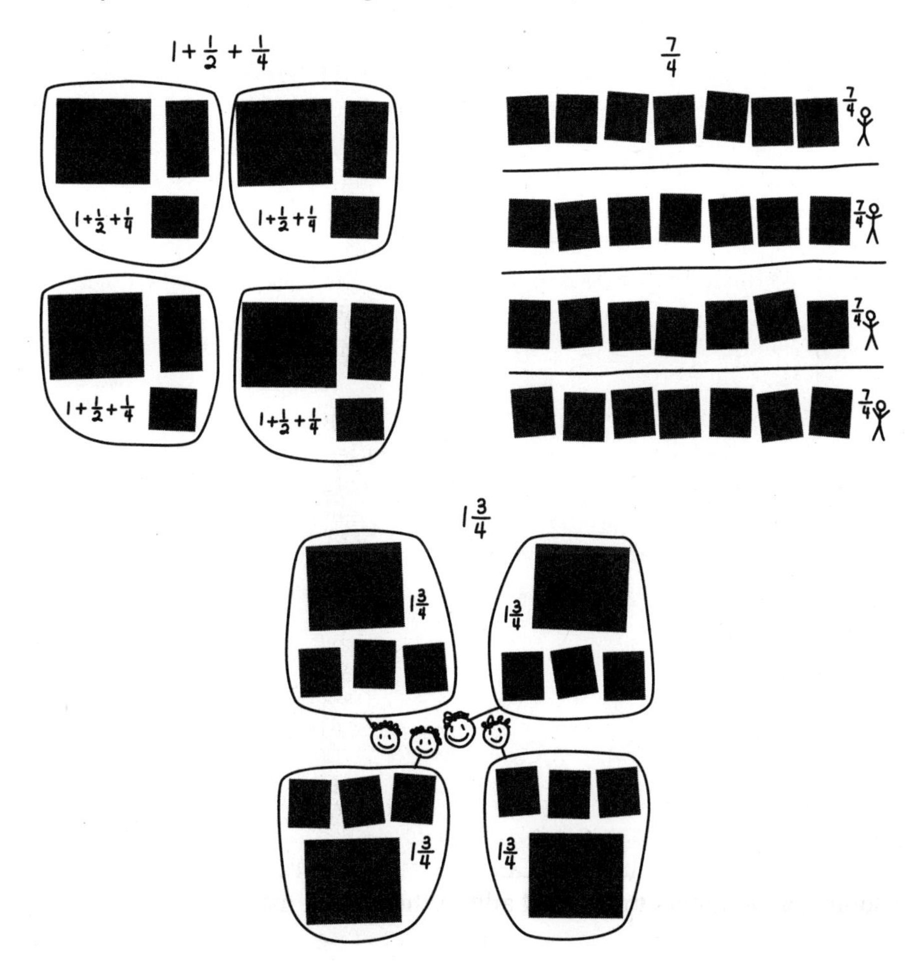

Write their solutions in fraction notation. When all answers are on the board, ask about the different names.

Are these answers the same? How can you show that they are?

Discuss any names for the share that are unclear to the students, and make necessary corrections.

Before tackling other problems, students might want to add ideas from this problem to the posted list of fraction facts. For example, they might reassemble the parts to make the whole:

$$\frac{7}{4} + \frac{7}{4} + \frac{7}{4} + \frac{7}{4} = 7$$

Or they might note that the different names for the shares are equal:

$$\frac{7}{4} = 1 + \frac{3}{4}$$

Introduce the notation for mixed numbers when you think this step is appropriate. Then this notation becomes:

$$\frac{7}{4} = 1 + \frac{3}{4} = 1\frac{3}{4}$$

Activity

More Sharing Problems

Distribute the extra copies of Student Sheet 3, Sharing Several Brownies, and pose other brownie problems, as suggested below. Students will continue to cut small brownies from the colored sheets to show their solutions.

You might first assign only problems that involve sharing among 2, 4, or 8 people, so that students work only with fractions with these denominators. Or you might mix the problems and let students puzzle through the differences.

How can 2 people share 3 brownies?
How can 2 people share 5 brownies?
How can 3 people share 4 brownies?
How can 3 people share 5 brownies?

How can 4 people share 2 brownies?
How can 4 people share 3 brownies?
How can 3 people share 2 brownies?
How can 6 people share 4 brownies?

The first few problems provide more than one brownie per person. However, in the second group of problems, each person receives less than a whole brownie. Write more than one problem on the board so that students can move on when they are ready.

Students can work in pairs at first, planning together how to solve the problems. However, they each divide their own brownies and paste their solutions on individual copies of Student Sheet 3. (They will need a new student sheet for each problem they do.)

As necessary, remind them that their task is not simply to divide the brownies in so many ways, but to show a *fair* sharing of the brownies. Students should be able to convince themselves and others that their approach is fair to everyone, as they will be asked to do in writing in the next activity.

Activity

Teacher Checkpoint

Writing About Shares

Students individually draw and write about their solutions in the preceding activity, More Sharing Problems, including a name for each person's share. If there is room, they may write on Student Sheet 3; otherwise, they can use notebook paper.

This activity gives you an opportunity to observe and assess students' strategies for solving these problems. As they work, circulate and ask students about what they are doing. You will also look over their work when they are finished.

Students will work at different paces. Some students may complete and write about only one or two problems in a session. Following are some things to consider in evaluating their work:

- Do students make equal shares?
- Do they maintain the whole, or do they (incorrectly) throw away some of the pieces?
- Can they describe a share with fractions in words?
- Can they write fraction notation to represent a share?

Reading Each Others' Solutions Students get together in small groups to read each others' solutions, both to see whether they make sense and to compare their answers. As in the first problem, students are likely to give different names for the same solution (for example, "one and a third and another third," "five of the 1/3 pieces," or "1 and 2/3"). When students agree that their solutions make sense, they list their answers on the board or on chart paper. Write column headings to help students group their solutions by problem. For example:

2 people, 3 brownies	4 people, 3 brownies
one and one-half	one half and half of a half
$1 + \frac{1}{2}$	$\frac{1}{2} + \frac{1}{4}$
$\frac{1}{2} + \frac{1}{2} + \frac{1}{2}$	$\frac{3}{4}$
$1\frac{1}{2}$	
$\frac{3}{2}$	

Have students save their completed work on all copies of Student Sheet 3, Sharing Several Brownies, in their folders. They will use them in Investigation 2 when they compare sizes of different shares.

Activity

Comparing Names of Shares

Toward the end of each session, pause briefly for students to compare and check the answers that they have written. Students check for themselves, perhaps talking with a partner, that all the answers to a given problem mean the same amount. They explore why answers that look or sound different actually mean the same thing. Discuss with the class any answers that students don't understand or can't reconcile with the other answers.

Hand out Student Sheet 4, Naming Fraction Shares, for students to begin independently in class. After they have had some time to work on problem 1, hold a brief discussion of their answers. Display the transparency of Student Sheet 4 as you invite students with different answers to explain their thinking. For a discussion of some of the issues raised by problem 1, see the **Teacher Note**, Maintaining a View of the Whole (p. 32).

Sessions 3 and 4 Follow-Up

Students can finish Student Sheet 4, Naming Fraction Shares, as homework.

Sharing 4 Brownies Among 5 People The following problem may be challenging for some students because it invites them to divide some of the brownies into quite small pieces that they will need to name with reference to the whole. Encourage students to find their own ways to solve it.

> PROBLEM: How can 5 people share 4 brownies equally?

Third grade students frequently invent the following strategy: Give each person a half, then a quarter. Divide the remaining quarter into 5/20. Give each person 1/20.

Write the solution this way:

$\frac{1}{2} + \frac{1}{4} + \frac{1}{20}$

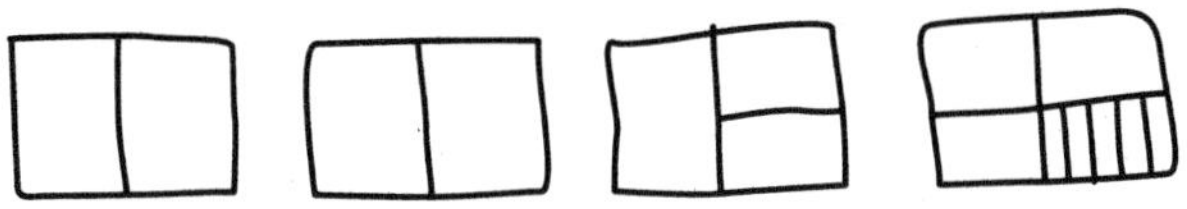

A few students may come up with this simpler strategy: Divide each brownie into fifths. Give each person four-fifths.

Teacher Note — *Maintaining a View of the Whole*

In doing cut-up brownie problems, third graders often lose sight of the whole. That is, they may cut a rectangle in half, assign that half to a person, and then cut the remaining half into three or four parts, calling these "thirds" or "fourths" instead of sixths or eighths. They forget the whole brownie-rectangle that the pieces came from.

Thus, for example, to divide two brownies among three people, a student might make this drawing, saying that each of the three people gets one-half and one-third. (The "third" they identify in this drawing is really a sixth.)

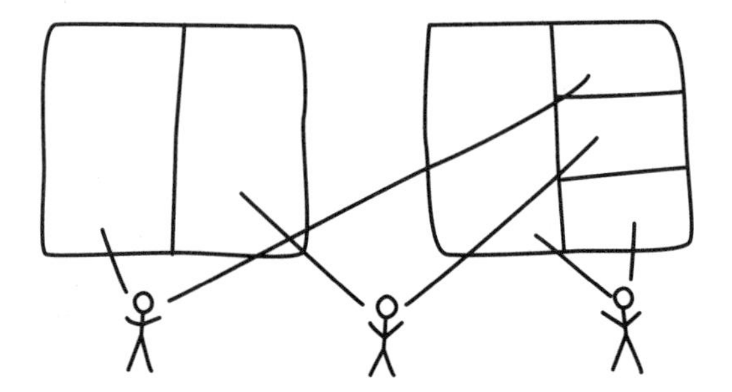

On Student Sheet 4, Naming Fraction Shares, problem 1 poses questions to help counter this confusion. In the first example, the share includes a half on one side of the rectangle and two-sixths on the other side, but even the half shows the divisions between the sixths (to counter the confusion of thirds with sixths that often arises because the sixths are thirds of a half).

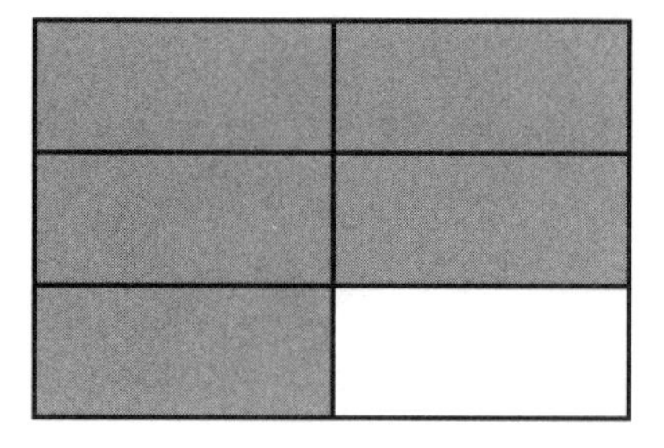

In the second example, the share includes a half on one side of a rectangle and one-eighth on the other side (one-fourth of the other half). The drawing does not mark the divisions between the eighths in the half, and so will be more challenging for the students than the previous example.

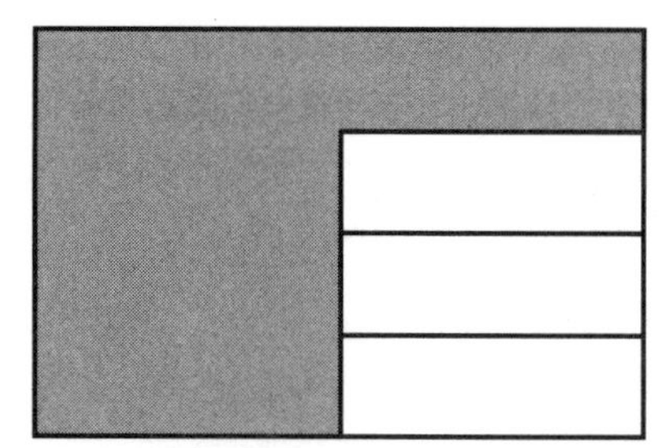

Although this student sheet is to be assigned at the end of the first investigation, plan class time to check over students' work and to discuss this issue with the whole class or with individual students who remain confused.

7 Brownies, 4 People

In this discussion of the Sharing Seven Brownies activity (p. 26), students talk their way through the dilemma of how to divide seven brownies equally among four people.

Amanda: First each person would get one.

Chantelle: But because there were only three left, and seven is an odd number—

Amanda: We give half to each person.

Chantelle: And have one left, and you can't divide it in half because there are four people.

Amanda: So we divide it into quarters and add a quarter to each half. That is one and a third.

This is what one-third looks like [*draws a whole rectangle and splits it into thirds*], and one and one-third would look like a whole plus one of these pieces.

Amanda: Oh. So what would half plus a fourth be?

That is the question! Let's draw what each person would get. Draw me one-half plus one-fourth.

Amanda [*Drawing a new picture in which she shades one-half and one-fourth in the same whole*]: Oh, it is three-fourths.

Chantelle: That makes sense, Amanda, because there are two-fourths in one-half, remember from yesterday?

Amanda: So each person gets one and a half and a quarter.

[*Teacher writes the following on the board:*]

one and a half and a quarter

$$1 + \frac{1}{2} + \frac{1}{4}$$

Does anyone have another way of sharing seven brownies among four people?

Jeremy: Give one whole to each person, then split the three brownies into four parts, one for each person.

Tell me what I should do to the picture.

Jeremy: Split the three brownies with a line this way [*makes vertical motion*], and this way [*makes horizontal motion*]. Now each person gets a piece from each brownie.

What size pieces are these?

Students: Quarters. Fourths.

So, how much brownie does each person get?

Jeremy: One whole brownie and three of the fourths pieces.

[*Teacher writes the following on the board:*]

one whole and three of the fourths

$$1\frac{3}{4}$$

Does anyone have another way?

Elena: You divide each of the brownies into four parts.

Yoshi: Yeah, that's how I was thinking about it. You give seven pieces to each person.

Elena: Each person gets seven of the fourths.

[*Teacher writes on the board:*]

$$\frac{1}{4} + \frac{1}{4} + \frac{1}{4} + \frac{1}{4} + \frac{1}{4} + \frac{1}{4} + \frac{1}{4} = \frac{7}{4}$$

We see through this discussion that different students organized the problem in different ways. Chantelle and Amanda gave each person one whole, one-half, and one-quarter. Jeremy gave each person one whole and three-quarters. And Elena and Yoshi gave each person seven-quarters. In each case, the teacher uses the students' descriptions of the pieces to write the answer using fraction notation.

Pattern-Block Cookies

What Happens

Sessions 1 and 2: Making Cookie Shares Students work in pairs to find and draw all the ways to make the equivalent of one yellow pattern block using blue diamonds (thirds), green triangles (sixths), and red trapezoids (halves). They write the fraction of the whole that each block represents. Students find the different shares of hexagon "cookies" they could give away and write some ways to make these shares.

Session 3: Comparing Shares Students discuss their homework on ways to make shares. Then they compare the sizes of cookie and brownie shares in different sharing situations. At the end of the session, students write a letter to a younger child explaining why one and one-third is a larger share than one and one-fourth.

Session 4: The Fraction Cookie Game Students play a pattern block and fraction dice game that involves adding and subtracting fractions. They also learn a game with their Fraction Cards.

Sessions 5 and 6: Backward Sharing Students use pattern blocks to solve problems that give the number of people and the size of each share. Then they solve problems that give only the size of the share. Students make lists of the number of people and the sizes of the shares, and they look for patterns.

Session 7 (Excursion): Half Yellow Students find ways to make pattern block designs that are half yellow. They draw, color, and label at least one design. After they make a design that is half yellow, some students may make designs containing other fractions of yellow. For homework, they write how they know their designs are half yellow.

Mathematical Emphasis

- Developing familiarity with common equivalents, especially relationships among halves, thirds, and sixths (for example, students exchange $2/6$ for $1/3$ and $3/6$ for $1/2$; they may also begin to make exchanges based on $1/6 + 1/3 = 1/2$ and $1/2 + 1/6 = 2/3$)
- Understanding that the relationships that occur between 0 and 1 also occur between any consecutive whole numbers (for example, $1/2 + 1/6 = 2/3$, so $2\,1/2 + 1/6 = 2\,2/3$)

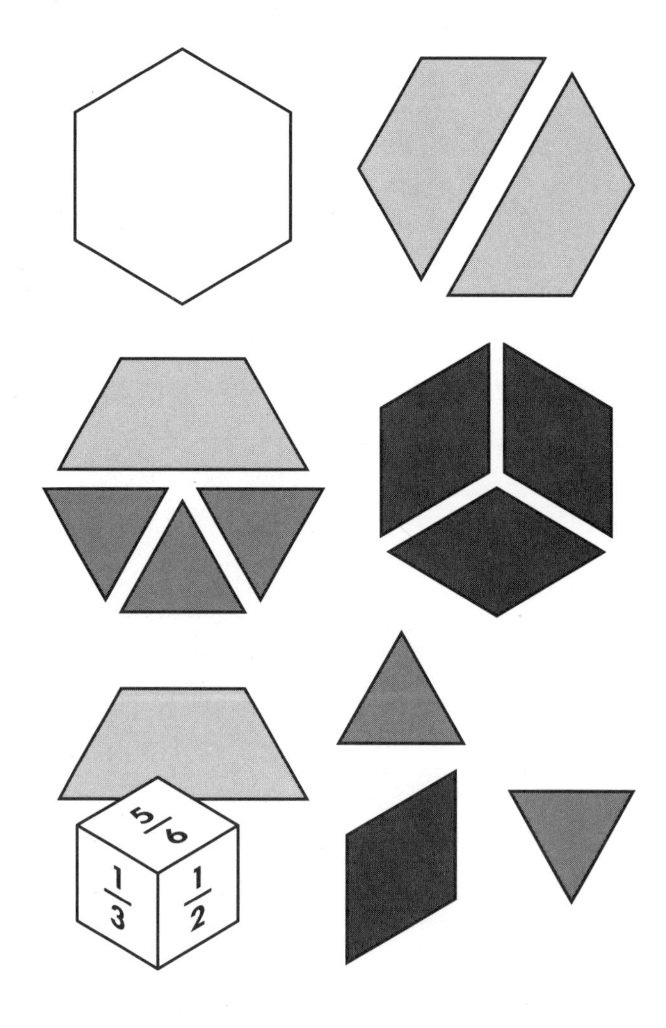

What to Plan Ahead of Time

Materials

- Pattern blocks: 1 bucket per 4–5 students (all sessions)
- Fraction dice in two colors: 3 per pair—2 in one color, 1 in the other color. Blank dice or inch cubes that you label with fractions can be substituted. (Session 4)
- Colored pencils, markers, or crayons (red, blue, green, and yellow) (all sessions)
- Classroom lists of fraction facts from Investigation 1 (Sessions 1–2)
- Overhead projector (Sessions 1–4)
- Calculators: 1 per student or pair (Ten-Minute Math)
- Pad of stick-on notes (Sessions 1–2 and 7)

Other Preparation

- Prepare the buckets of pattern blocks by removing the orange squares and tan rhombuses (thin diamonds); students will be using only the red trapezoids, blue diamonds, yellow hexagons, and green triangles.
- Familiarize yourself with the fractional relationships among pattern block pieces.
- If you do not have manufactured fraction dice, label the faces of blank dice or inch cubes with these fractions:

  $\frac{1}{2}$ $\frac{1}{2}$ $\frac{1}{3}$ $\frac{2}{3}$ $\frac{1}{6}$ $\frac{5}{6}$

- Duplicate student sheets and teaching resources (located at the end of the unit) in the following quantities:

  For Sessions 1–2

  Student Sheet 5, Hexagon Cookies: 4–5 per student (save 2–3 for Session 4), plus 1 transparency

  Student Sheet 6, Many Ways to Make a Share: 1 per student (homework)

  For Session 3

  Student Sheet 7, Who Gets the Larger Share? (2 pages): 1 per student, plus 1 transparency

  Student Sheet 8, Letter to a Second Grader: 1 per student

  For Session 4

  Student Sheet 9, How to Play the Fraction Card Game: 1 per student

  For Sessions 5–6

  Student Sheet 10, How Many Altogether? 1 per student

  For Session 7

  Triangle Paper: 2–3 sheets per student

Sessions 1 and 2

Making Cookie Shares

Materials

- Pattern blocks
- Student Sheet 5 (2 per student)
- Transparency of Student Sheet 5
- Transparency pens in red, blue, yellow, and green
- Lists of fraction facts (from Investigation 1)
- Student Sheet 6 (1 per student)
- Overhead projector
- Colored pencils, markers, or crayons
- Stick-on notes

What Happens

Students work in pairs to find and draw all the ways to make the equivalent of one yellow pattern block using blue diamonds (thirds), green triangles (sixths), and red trapezoids (halves). They write the fraction of the whole that each block represents. Students find the different shares of hexagon "cookies" they could give away and write some ways to make these shares. Their work focuses on:

- finding shapes to cover one whole
- identifying fractional parts that add to one whole
- writing fraction expressions

Ten-Minute Math: Broken Calculator Once or twice during the next few days, in a spare 10 minutes, play Broken Calculator. Pairs of students can share a calculator, but the activity is better if each student has one. Focus the activity on division problems that give targeted whole-number answers.

Choose a target number for students to get on their calculator display. Begin with a small number, perhaps 2 or 5 or 10. (Students will find interesting patterns when they try to get 10.)

How can we use division to get an answer of 2 (or 5 or 10 or a larger number) on the calculator? Pretend that the plus, minus, and multiplication keys are broken.

Give students time to find ways to get the number. They can talk together as they try. Write any ways that students find on the board so that other students can try them. Discuss what all the answers have in common. Give students a chance to find more solutions.

For full directions and variations on Broken Calculator, see p. 74.

Activity

Cutting Up Cookies

On an overhead projector, display the transparency of Student Sheet 5, Hexagon Cookies. Have some pattern blocks available (yellow, red, blue, and green) to cover the hexagons on the transparency.

Let's say that this yellow pattern block is a cookie. How could we make a cookie like this with other pieces?

How many blue pieces do we need to cover a yellow cookie completely? How many red pieces? green pieces?

I'm going to give you a sheet filled with pictures of pattern-block cookies. Your task is to find lots of different combinations of red, blue, and green pieces that will exactly cover a yellow hexagon cookie.

Hand out Student Sheet 5, one to each student. They use pattern blocks to find solutions, then record their solutions on this sheet by drawing the smaller pieces onto the hexagons. For an orderly approach that will help them avoid repetition, they might start by finding all the ways of covering a yellow cookie with just one other color, and then use just two colors, and finally use three colors. However they approach the problem, they should check their work to be sure they have no repeats.

Leaving the Hexagon Cookies transparency on the overhead or passing it around the room, ask students to draw the different ways they found to make the hexagon with the other pattern blocks. (You might want to begin with a few students who tend to be slow starters, then have some other students add one solution each.) If you have red, blue, and green transparency pens, let students use those to indicate the corresponding pattern blocks.

Display the completed transparency on the overhead. Students may copy any solutions that they don't have onto their own sheets, and check for repeats or missing solutions both on their papers and on the transparency. There are seven different ways to cover the yellow hexagon "cookie" with two or more pieces (see student's sample answers, p. 38).

Naming the Fractional Parts When students have drawn all the ways to divide the hexagon they can find, they write the corresponding fractions in each of the pieces. You might first have students identify a piece that is *half* the yellow cookie. When students agree on a piece that is half of the whole, write the fraction $1/2$ in some of the red trapezoids on the transparency. Students then work on their own to decide what fractions the pieces of each hexagon represent, and write those fractions in all the corresponding spaces.

As you observe students, ask them how they are figuring out the fractional parts and encourage them to explain their methods to other students. If students are stuck, suggest that they make hexagons with only one color. Then ask:

How many pieces did you use to make the whole? What fraction is each piece?

As students finish, ask them to write number sentences next to the hexagons to show what fractions they used to make one whole, as the student has done below.

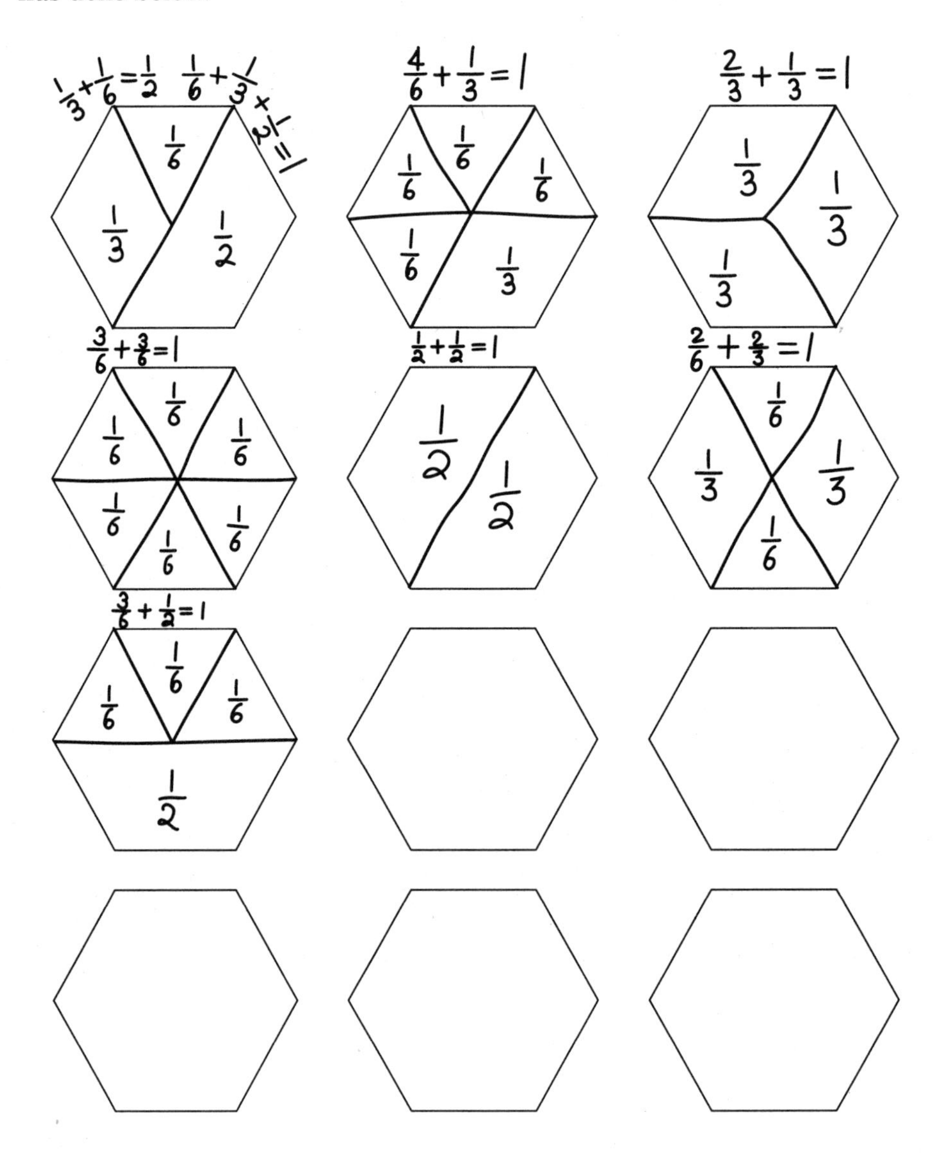

What Fractions Can You Give Away?

We found many ways of making a whole yellow cookie. What are some of the fraction sentences that you wrote next to your hexagons?

As students tell you different ways of making one whole, write the solutions as number sentences on the fraction facts list. Begin a section for facts about thirds, sixths, and halves if you haven't already.

Then ask students to consider how they might give away parts of pattern-block cookies.

You could give away a whole cookie in several ways. Let's figure out what smaller fractions of this cookie we could give as one share.

Could we give away one-half a cookie? What are the ways you could make one-half of a cookie? How would you write those ways in number sentences?

Add a few of the students' suggestions to the list of fraction facts.

Could you give away one-fourth of a yellow cookie?

Students will discover that they can't give away one-fourth using this set of blocks. Some students may use two green blocks (two-sixths) and two blue blocks (two-thirds) as four unequal parts of a yellow cookie. Point out that because these shapes are unequal sizes, they are not fourths. Ask students what shape they would need in the pattern blocks to make fourths. (They would need a trapezoid that divides the hexagon into equal fourths; see if any students can draw this on the transparency.)

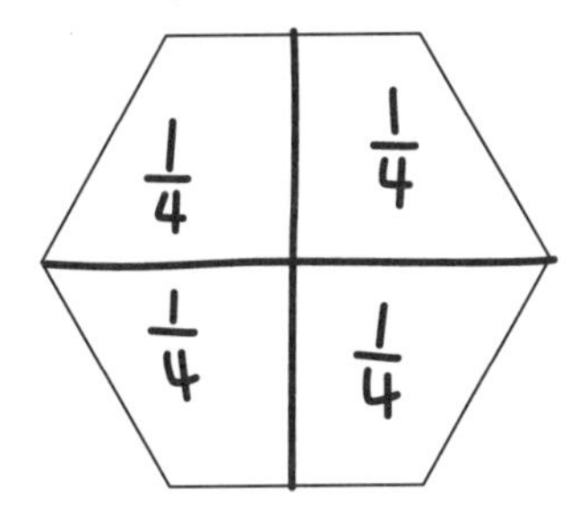

Could you give away one-third of a yellow cookie? (Yes, as one blue diamond or two green triangles.) **What are all the parts of a yellow cookie that you could give away?**

Students work with partners to list all the different-sized shares they could give away (1, $1/2$, $1/3$, $1/6$, $2/3$, $5/6$). This task should take only a few minutes. You might circulate and observe their lists while you hand out Student Sheet 6 for the next activity.

Activity

Making Shares in Many Ways

Introduce Student Sheet 6, Many Ways to Make a Share, at the board. Write one of the headings from the sheet, and ask students for their ideas.

How can we make one-half from smaller pieces? Can you think of another way? Think about hexagon cookies. What pieces can you use to make half of a hexagon cookie? Then think about our paper brownies. How can you make half a brownie? How would we write that in fractions?

As students make suggestions, write their ideas on the board in fraction notation.

How can we make one-third from smaller pieces? How can we make two-thirds?

Once students have the idea, they can start working independently or with partners on Student Sheet 6, writing expressions under each of the different headings. Although they start the sheet in class, they will continue to add more expressions as homework.

To make some of the fractions larger than $1/2$, challenge students to combine fourths and sixths or eighths and sixths (for example, one way to make $3/4$ would be $3/6 + 1/4$).

Activity

More Fraction Facts

During this class or at a later time, work with small groups or the whole class to expand the lists of fraction facts that you started in Investigation 1. Most of the facts from sharing brownies in the first investigation involved halves, fourths, and eighths; most of the facts from the pattern-block cookies involve thirds and sixths. Also ask students if they have found any surprising combinations at home working with Fraction Cards. You might start a new list of facts that combine fractions from the first two lists.

After listing facts students have observed in recent work, challenge students to make number sentences that join fraction combinations representing equal amounts. For example:

1 = 2 halves	$\frac{1}{3} + \frac{1}{3} + \frac{1}{3} = 1$, so $\frac{1}{2} + \frac{1}{2} = \frac{1}{3} + \frac{1}{3} + \frac{1}{3}$
$\frac{2}{3} = \frac{1}{3} + \frac{1}{3}$	$\frac{2}{3} = \frac{1}{6} + \frac{1}{6} + \frac{1}{3}$, so $\frac{1}{3} + \frac{1}{3} = \frac{1}{6} + \frac{1}{6} + \frac{1}{3}$

Students can add more facts to the list as they discover them. Make stick-on notes available for students to initial and place near facts they don't understand or that they think are incorrect. When some facts are in question, take time out for the students to discuss them.

Sessions 1 and 2 Follow-Up

Students continue to add solutions to Student Sheet 6. Encourage them to get ideas from family members, but to be sure to write only those ideas that make sense to them.

Session 3

Comparing Shares

Materials

- Completed Student Sheet 6 (from homework)
- Student Sheet 7 (two pages, 1 set per student)
- Transparencies of Student Sheet 7
- Completed work on Student Sheet 3 (saved from Investigation 1)
- Overhead projector
- Student Sheet 8 (1 per student)

What Happens

Students discuss their homework on ways to make shares. Then they compare the sizes of cookie and brownie shares in different sharing situations. At the end of the session, students write a letter to a younger child explaining why one and one-third is a larger share than one and one-fourth. Their work focuses on:

- working with fractions and mixed numbers
- determining fair shares
- comparing shares
- proving which fraction is larger

Activity

Reviewing Homework

Take some time for students to discuss results from their homework on Student Sheet 6, Many Ways to Make a Share. Have them form groups of three or four to compare their answers and explain their thinking. The group together decides which answers make sense and which need rethinking.

After 5 to 10 minutes, call students together and ask a few students to tell what they learned from the discussion. Ask them to comment only on their own thinking, not on someone else's errors. Encourage students to keep Student Sheet 6 in their folders and to add new combinations to it whenever they get new ideas.

Activity

Which Is More?

For this activity, students work on Student Sheet 7, Who Gets the Larger Share? (two pages). The sheet presents four problems, asking students to compare ways of sharing brownies (on the first page) and ways of sharing hexagonal cookies (on the second page). Students decide in each case which group of people will receive the larger share.

Students may consult with their neighbors, but each completes his or her own copy of the student sheet. They can use their completed work on Student Sheet 3, Sharing Several Brownies, where they may have already worked out equal shares for some or all these problems.

The main task for students is deciding which group gets the larger share, and explaining their thinking in solving the problem. Some students might draw a picture showing each share in order to decide which group gets more, but they don't necessarily have to draw first. There are many ways to make the decision. See p. 44 for some ways of reasoning about these problems.

Encourage students to think *before* using drawings. For example, in the first problem, students should be able to recognize with some good thinking that two people will get a larger share than three people, because fewer people are sharing the same amount.

Discussing Results When everyone has finished both pages of Student Sheet 7, bring the class together to share results. Display the transparencies of the student sheet, one page at a time. For each problem, poll the students to find out which share they believe is larger. Ask students to explain their choices.

As the students tell you how large each share is, write this information on the transparency. Write each share in several ways for ease in making comparisons (for example, include both $1\frac{2}{3}$ and $\frac{5}{3}$). If there are differences of opinion, students discuss the problems with their neighbors until they agree about which share is larger, or that the shares are the same size.

The answers to Student Sheet 7 are as follows:

1. Five brownies for 2 people is a larger share than 5 brownies for 3 people, because fewer people share the same number of brownies. The 2 people each get $2\frac{1}{2}$ brownies, as compared to $1\frac{2}{3}$ brownies for the 3 people.
2. The shares are the same size in the two groups. The first sharing simply provides twice as much brownie for twice the number of people. Each person gets $\frac{4}{6}$ or $\frac{2}{3}$ of a brownie.
3. Three cookies for 2 people is a larger share than 4 cookies for 3 people. In both cases they each get one whole cookie plus part of one extra cookie; the difference comes in sharing the extra cookie among 2 people instead of 3 people. Thus, the 2 people get $1\frac{1}{2}$ cookies each, as compared to $1\frac{1}{3}$ for the 3 people.
4. Two people sharing 7 cookies leaves a larger share per person than 3 people sharing 8 cookies. The two people each get $3\frac{1}{2}$ cookies, as compared to the $2\frac{2}{3}$ cookies that the 3 people get.

Activity

Assessment

Letter to a Second Grader

For this assessment task, students work on Student Sheet 8, Letter to a Second Grader. They are asked to decide which amount is larger, $1\frac{1}{4}$ or $1\frac{1}{3}$, and to write an explanation of their answer that a second grader could understand.

❖ **Tip for the Linguistically Diverse Classroom** Give students the option of writing their letters in their native languages. If they cannot write in their native language, most of their ideas can be conveyed through number sentences (1 and $\frac{1}{4}$ < 1 and $\frac{1}{3}$, or vice versa) and simple drawings.

Students will need about 20 minutes of class time to do this task. Collect their work and examine it to evaluate their level of understanding.

In evaluating student work, consider what arguments the students used to convince someone else, and how persuasive they are. Students who believe that $\frac{1}{4}$ is larger than $\frac{1}{3}$ may make drawings that support this error. For example, they may draw each fourth the same size as each third so that the 4 fourths make a larger whole than the 3 thirds. Then they may argue that "fourths are larger than thirds."

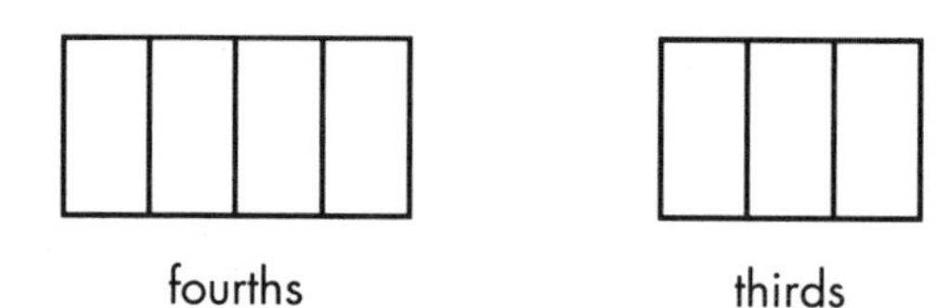

The most common correct explanation will have a drawing in which the wholes are the same size, so that the fourths are smaller.

Watch for students who make correct drawings and give correct answers, but betray a lack of understanding in their explanation. Such a student might argue, "1/3 is bigger because only two pieces are left over. When you take 1/4, three pieces are left over." By such an argument, 1/2 would be larger than 8/10. A more generally correct answer describes the size of the share in wholes of the same size: "I think 1/3 is bigger because it takes up more room."

An even better answer will explain that if more people share the same amount, each person gets a smaller piece.

Name Date

Student Sheet 8

Letter to a Second Grader

Some people think 1 and $\frac{1}{4}$ is a larger share than 1 and $\frac{1}{3}$.

Some people think 1 and $\frac{1}{3}$ is a larger share than 1 and $\frac{1}{4}$.

Which share do you think is bigger?
Write a letter to a second grader. Tell why you are right.
Use drawings to explain your thinking.
Remember, a second grader must understand what you write.

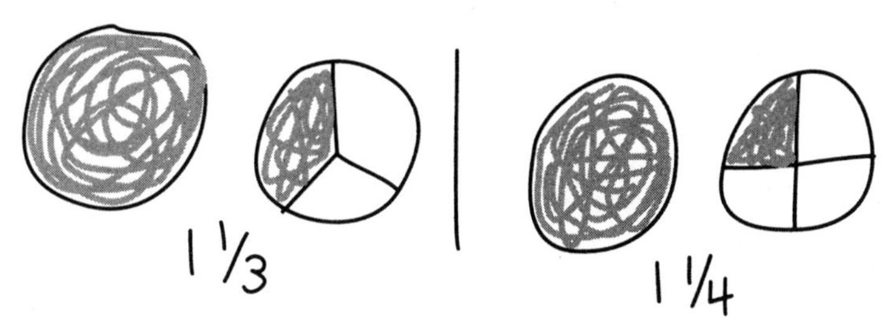

If you had to cut Something into 4 peices there would be more of them So the peices would be smaller then if you cut them into 3's.

Session 4

The Fraction Cookie Game

Materials

- Fraction dice in two colors (3 per pair)
- Pattern blocks
- Student Sheet 5 (2–3 per student)
- Colored pencils, markers, or crayons
- Student Sheet 9 (1 per student)
- Fraction Cards students' sets from Investigation 1)
- Overhead projector

What Happens

Students play a pattern block and fraction dice game that involves adding and subtracting fractions. They also learn a game with their Fraction Cards. Their work focuses on:

- identifying fraction parts
- exchanging equivalent fractions
- adding and subtracting fractions

Activity

Game Warm-Up

Introduce the Fraction Cookie game to the whole class. Explain that in this game, students are to collect pieces of pattern-block cookies. The basic game works with one fraction die. Players will take turns rolling the die and then add to their cookies the exact amount of their roll. Demonstrate briefly by rolling one fraction die and putting the pattern block representing that fraction on the overhead. Roll again, and add the new block or blocks to your display.

Trading Up Another basic part of the game is "trading up." Players must trade two or three pieces for a larger piece whenever possible, so that at the end of each turn their total number of collected cookies is represented with the fewest possible pattern blocks. For example, a player with $2\frac{1}{2}$ cookies should have two yellow cookies and one red half at the end of the turn. Players should be alert to ways they could combine greens, or a green and a blue, to make larger pieces. Explain and demonstrate this on the overhead.

Each player starts with a blank copy of the Hexagon Cookies sheet (Student Sheet 5) on which to place the collected pieces. If yellow hexagon blocks are in short supply, students may use the sheet for recording their completed cookies; each time a cookie is complete, the player can remove the blocks from the sheet and color in the cookie. See the **Dialogue Box**, Playing the Fraction Cookie Game (p. 49), for a portion of a game between two students.

Activity

Playing the Fraction Cookie Game

Students play in pairs. When they seem to tire of the beginning version (for some, this may happen quite quickly), introduce the two more challenging versions explained below. Some students will play the simpler versions longer than others; there is no need to rush them to a higher level.

Beginning Game (One Die) Players roll one die and add the amount on the die to their cookie collections. For example, if the die shows 1/3, the player takes a blue diamond (1/3 of a yellow cookie). Each player completes a turn without interruption; the other player then checks the first player's work. Encourage overly helpful students to give quieter students a chance to decide for themselves how they want to play out their turns and make their trades. Let students decide how many cookies a player needs to win the game.

Intermediate Game: Adding Fractions (Two Dice) As pairs of students seem ready, hand out another die so that they can throw two dice and add the fractions to determine how much cookie to take. As in the beginning game, players finish their turn by making the trades necessary to get the fewest pieces in their cookie collection. Partners check each other's work after each turn.

Advanced Game: Adding and Subtracting (Three Dice) When students are comfortable playing with two dice, introduce a third die of a different color. On each turn, players roll two dice of one color and a third die of a different color. They add the amounts on the first two dice and subtract the amount on the third die from their cookie collections.

In this version of the game, players start with two hexagon cookies (so they won't run out when they subtract). The first player to get four cookies (or students can decide on another number) wins. Some students will be ready to try this variation before others.

Activity

The Fraction Card Game

Toward the end of the session, hand out copies of Student Sheet 9, How to Play the Fraction Card Game. To play the game in class, pair up students who have Fraction Card sets in different colors, or have them mark their sets with initials or a colored marker so they can tell the two sets apart. This game requires some space, so try to play on the floor or a clear table.

The two players each mix up their own Fraction Cards and stack them, number side down, in front of them. At the same time, the two players turn over one card from the top of their set, so that the fraction label shows. The player who puts down the larger Fraction Card takes both cards. If the cards are the same size, each player turns over another card, and the player with the largest card takes all four. (To this point, the game is much like the card game War, which students may already know.)

As players win cards, they must try to make a "whole" with them, putting the fractional parts together. (This is what takes up space!) Any cards that the player has already won may be used at any time to form a whole.

Play continues until all the cards have been turned up and won by either player. The person with the most wholes at the end is the winner.

Allow class time for one or two sample rounds.

Session 4 Follow-Up

Students take home their class set of Fraction Cards and Student Sheet 9, How to Play the Fraction Card Game. They should already have another set of Fraction Cards at home (made as homework during Investigation 1), so they can play the game with friends and family members. As necessary, send home extra copies of How to Make Fraction Cards (p. 96).

Playing the Fraction Cookie Game

Annie and Ricardo are playing at the beginning level of the Fraction Cookie game (p. 47). This sample dialogue shows how they deal with trading up, and with a shortage of yellow pattern blocks.

Annie: I rolled 1/6, so I need a green.

Ricardo: I got 1/2. OK, so I'll add a red. And I can trade in my two reds for a yellow.

Annie: OK, I'll go again. I got 1/3. That is a blue. Hmm, can I do any trades? I could trade my green and blue for a red.

Ricardo: If you had two blues, you could trade them for four greens.

Annie: But I don't want more pieces, I want less.

Ricardo [*as he rolls*]: Let's see ... 2/3, that's two blues. I'm trading two blues and two greens for two reds.

Annie: Just make it a yellow.

Ricardo: Oh yeah! But we ran out of yellows. I'll just color in a yellow, and then I'll know I made a whole.

Sessions 5 and 6

Backward Sharing

Materials

- Pattern blocks
- Student Sheet 10 (1 per student)

What Happens

Students use pattern blocks to solve problems that give the number of people and the size of each share. Then they solve problems that give only the size of the share. Students make lists of the number of people and the sizes of the shares, and they look for patterns. Their work focuses on:

- putting pieces together to make wholes
- looking for number patterns

Activity

How Many Cookies in All?

This activity is another variation on the brownie-sharing problems. Students are given the number of people who received shares and the size of each person's share, and must then find out how many brownies (or other objects) there were to start with. This type of "backward sharing" requires students to think about the fractional parts in relation to a whole that they must construct.

Student Sheet 10, How Many Altogether? presents a few of these story problems. Students can use pattern blocks as they solve them. Let students work individually on the first problem, which asks how many cookies Isaac has to bake so that he and five friends each have a share of $1\frac{1}{3}$ cookies.

❖ **Tip for the Linguistically Diverse Classroom** Read the first problem aloud, drawing on the board and using mime to make it comprehensible for all students. For example, smack your lips over a "pattern-block cookie," draw six stick-children in party hats, and draw 1 and $\frac{1}{3}$ cookies on a plate, connected by arrows to each of the six stick figures.

For students who can't think how to begin, you might suggest that they lay out each person's share in pattern blocks. Then they can combine parts to find out how many whole cookies Isaac had to bake.

After a few minutes, stop everyone for a brief discussion. Invite two or three students to explain their approaches for solving it. Students then continue with the next two problems. As you observe them working, ask them to show clearly how they are doing the problem so you can understand their thinking when you read their work later.

Those who have time can begin to work on the Challenge problem at the bottom of the student sheet, but you will be introducing this special problem to the whole group in the next activity.

Activity

How Many People? How Many Cookies?

The Challenge problem on Student Sheet 10 is more open-ended, because it gives only the size of the share ($1\frac{1}{2}$ cookies). Students must determine both how many cookies the group started with and how many people might be in the group. They use pattern blocks to find many possible solutions. As students suggest answers, have them listed on the board in a chart.

Students work until they have several answers and begin to realize that the list could grow very long. They look for patterns in the numbers. For example, the number of people are multiples of 2, while the numbers of cookies are multiples of 3. Some students may even see that the number of cookies is always $1\frac{1}{2}$ times the number of people.

Write more such problems on the board for students to do on the back of Student Sheet 10. Make the shares familiar fractions. For example:

> How many people and how many whole cookies ...
>
> so that everyone gets $2\frac{1}{2}$ cookies?
>
> so that everyone gets $\frac{2}{3}$ of a cookie?
>
> so that everyone gets $1\frac{1}{3}$ cookies?

Collect lists of answers for each of these problems on the board. Have a brief discussion about patterns that show up in the charts.

In these problems, we always start with whole cookies. Why do you think only an even number of people get shares with halves?

For additional challenges, students can make up their own examples. Ask them to predict how many cookies would be needed for 20 or more people, without counting all the way.

Share = $1\frac{1}{2}$ cookies		Share = $2\frac{1}{2}$ cookies		Share = $\frac{2}{3}$ cookies		Share = $1\frac{1}{3}$ cookies	
People	cookies	People	cookies	People	cookies	People	cookies
2	3	2	5	3	2	3	4
4	6	4	10	6	4	6	8
6	9	6	15	9	6	9	12

Session 7 (Excursion)

Half Yellow

Materials

- Pattern blocks
- Colored pencils, markers, or crayons
- Triangle paper (2–3 sheets per student)
- Stick-on notes

What Happens

Students find ways to make pattern block designs that are half yellow. They draw, color, and label at least one design. After they make a design that is half yellow, some students may make designs for other fractions of yellow. For homework, they write how they know their designs are half yellow. Their work focuses on:

- creating a design that is half yellow or another fraction of yellow
- drawing and coloring a design
- recognizing that a design is half or another fraction yellow

Activity

Designs That Are Half Yellow

Working alone or with a partner, students use pattern blocks to make a design that is half yellow. For a discussion of some of the thinking connected with this problem and some confusions that may arise in your class, see the **Teacher Note**, Designs That Are Half Yellow (p. 57).

While students are making their designs with the blocks, observe their work and ask about their strategies.

How do you know that exactly half your design is yellow? How did you decide which blocks to use? What fraction of the design is not yellow?

After making two or three designs that are half yellow, students select their favorite design (or make a new one) to draw and color. Give them triangle paper to simplify this task.

Activity

Other Fractions of Yellow

For students who are ready for a greater challenge, pose this problem:

Make a design that is one-third yellow. How much of the design will *not* be yellow?

Extend the challenge for students who are ready by posing new problems that involve making designs with different fractions of yellow. For example, challenge them to make a design that is two-thirds, one-fourth, three-fourths, or one-sixth yellow. In each case, also ask how much is not yellow. Adjust the problems for individual students. For most students, working with halves will be enough of a challenge.

If your students need an introduction to showing fractions other than one-half, make a simple design of one yellow hexagon and three hexagons of a mixture of the other colors.

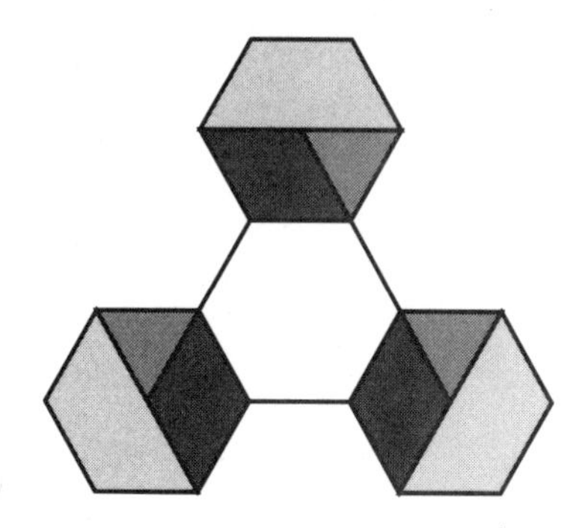

Ask how much of the design is yellow [one-fourth] and how much is not yellow [three-fourths]. Make your design on a tray you can carry around to show groups, or leave it on a table where students can go up in small groups to look at it, figure out the fraction of yellow, and whisper it to you or write it to show you. While students watch, move the non-yellow pieces around and ask again about the fraction of yellow. (It will be the same.)

Activity

How Do You Know It Is Half Yellow?

Finish up this session with a writing activity in which students describe one or more of their designs.

Pick one of your designs. Explain in writing how you know what fraction of your picture is yellow. Use the drawing of your design in your proof.

Students might start their explanation with this sentence:

I know this design is ____ [*fill in the fraction*] yellow because ...

You may need to model this approach by making your own design and describing how you know what fraction is yellow.

❖ **Tip for the Linguistically Diverse Classroom** Students can write their explanations in their native languages, or can explain how they know by labeling the different colored parts of their drawings with fractional notation.

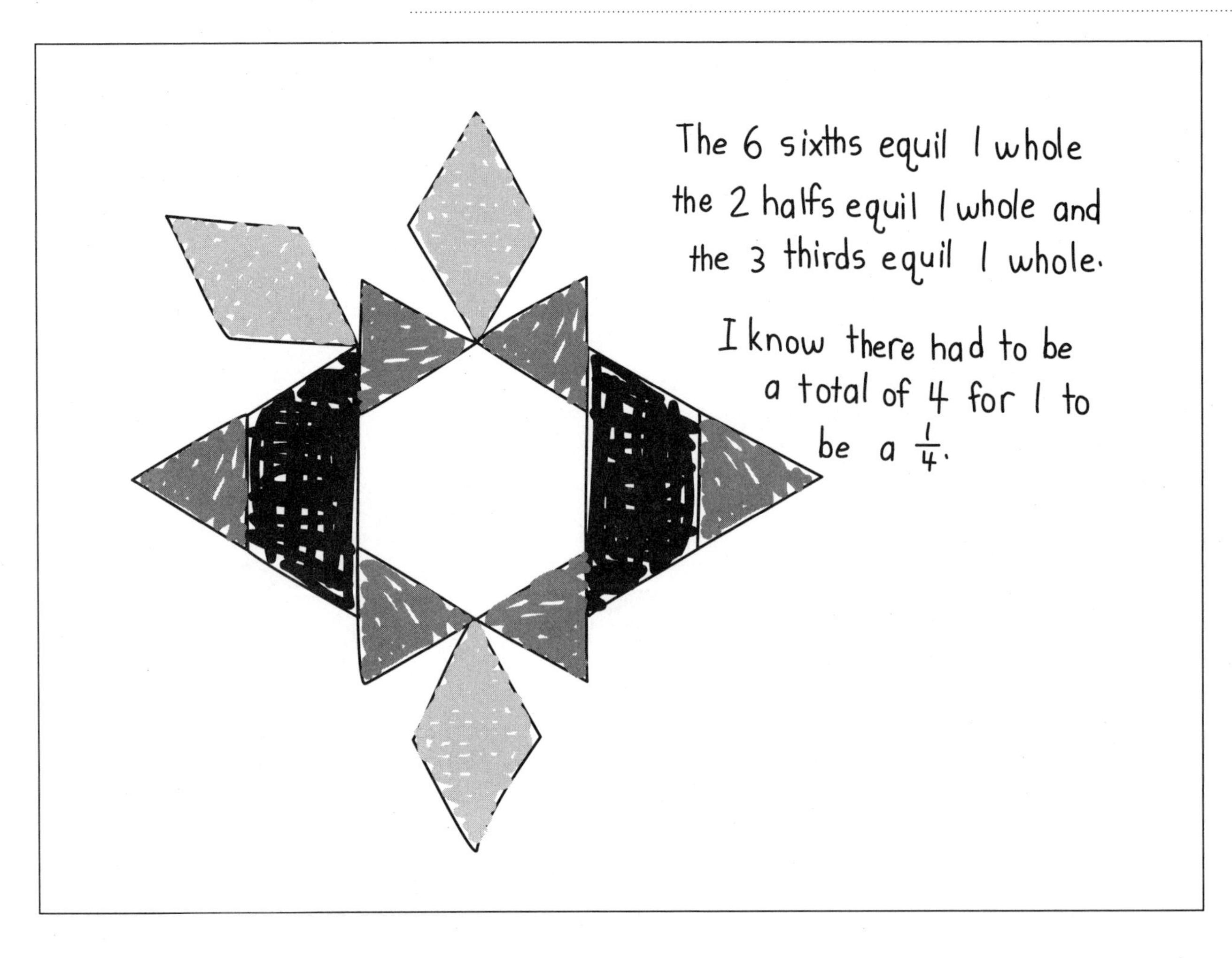

All the greens equal $\frac{1}{4}$. All the red equal $\frac{2}{4}$. The 1 yellow equals $\frac{1}{4}$ yellow.

$$\begin{array}{r} \frac{1}{4} \\ +\frac{2}{4} \\ +\frac{1}{4} \\ \hline \frac{4}{4} \end{array}$$

$\frac{1}{4}$

all of the green and the red equal $\frac{3}{4}$ and the yellow equal one $\frac{1}{4}$.

Session 7 Follow-Up

Which Fraction Is Yellow? If students make designs with a variety of fractions of yellow, use their designs as a puzzle for the class. Collect all the designs. Cover any specific reference on the design that indicates what fraction is yellow. Mix up the designs and give out a few to each group of four or five students.

Students put stick-on notes on each design, writing their initials and the fraction they think is yellow. Students who agree about the fraction can all write their initials on the same note.

Post all of the designs with the stick-on notes. Allow students time to look at the designs for which there are differing opinions. Finally, for these designs, have the creators tell the answer and explain how they know what fraction the yellow portion represents.

Designs That Are Half Yellow

Teacher Note

Making a design that has a certain fraction of one color is a more difficult task than breaking an already-drawn shape into equal parts. As students add new blocks to their designs, the size of the whole changes, and the size of each fractional part needs to change in proportion to the whole. Experiment with this exercise yourself to discover the difficulties.

One common confusion is to represent the numerator of the fraction with yellow hexagons and the denominator with hexagons of the other colors. For example, a student might try to make "1/2 yellow" using one yellow hexagon and two hexagons of other colors, and "2/3 yellow" with two yellow blocks and three of the other colors.

Or, students might think that half the *number* of blocks should be yellow. Because the other blocks are smaller than the yellow, they will need more of them—students need to focus on size (area) rather than number of blocks. To help students think through this confusion, ask them to pretend that their pattern is cookies. Is it all right for you to take half by taking all the yellows? Be sure they understand that the area needs to be half yellow.

Possible strategies include these two, which approach the problem from different angles:

- Choose the blocks first, and then make the design. One approach is to make hexagon towers—one yellow, and another that combines red, blue, and green blocks. When the towers are the same height, you know one-half the block area is yellow. Then use the blocks from the towers to make a flat design.

  This strategy also works well for making designs with other fractions of yellow. To make a design that's 1/3 yellow, make one yellow tower and two towers of the other colors. Then use all of the blocks from the three towers to make a design.
- Make a design, and then figure out if too much or too little of it is yellow. Modify the design as needed, taking away or adding blocks until the design is half yellow and half other colors.

INVESTIGATION 3

Other Things to Share

What Happens

Sessions 1 and 2: How Can We Split Balloons? Students think about how to share different kinds of items, including items that cannot be split into parts (balloons) and items that have a limited number of parts (money). To become familiar with decimal equivalents for common fractions, they find solutions to brownie problems with and without the calculator.

Session 3: Sharing Many Things Students determine how to share a large number of items equally with their group. They determine what fraction of the items each student should get, and then figure out how many of the items the fraction represents.

Mathematical Emphasis

- Understanding the relationship between fractions and division (for example, by solving problems in which the whole is a number of things rather than a single thing, and the fractional part of this whole is a group of things as well, as in $1/3$ of 6 is 2)
- Relating notation for common fractions ($1/2$, $1/4$, $3/4$, $1/5$, $1/10$) with notation for decimals on the calculator (0.5, 0.25, 0.75, 0.2, 0.1)
- Using different notations for the same problem (for example, $6 \div 2$ and $1/2$ of 6)

What to Plan Ahead of Time

Materials

- Calculators: 1 per student (all sessions)
- Play money in dollars, quarters, and dimes (Sessions 1–2, optional)
- Pattern blocks: 1 bucket per 4–5 students (Sessions 1–2)
- Small items to share, such as pennies, paper clips, or peanuts: 8–10 per student (Session 3)

Other Preparation

- Duplicate student sheets (located at the end of the unit) in the following quantities:

  For Sessions 1–2

  Student Sheet 11, Other Things to Share: 1 per student

  Student Sheet 12, Sharing With and Without a Calculator (2 pages): 1 per student

Sessions 1 and 2

How Can We Split Balloons?

Materials

- Student Sheet 11 (1 per student)
- Student Sheet 12 (1 per student)
- Calculators (1 per student)
- Play money (optional)
- Pattern blocks

What Happens

Students think about how to share different kinds of items, including items that cannot be split into parts (balloons) and items that have a limited number of parts (money). To become familiar with decimal equivalents for common fractions, they find solutions to brownie problems with and without the calculator. Their work focuses on:

- relating meanings for division and fractions
- relating notation for money to decimal notation on the calculator

Ten-Minute Math: Broken Calculator Once or twice during the next three days, play Broken Calculator. Each student needs a calculator for this activity. Pose a problem that involves division and gives a decimal answer.

Pretend the plus, minus, multiplication, and decimal point keys on your calculator are broken. How can you get 0.5 for an answer if you use only division? Work with your neighbors to find several ways.

Students who manage to get 0.5 several ways can then try to get 0.25 and 0.75 using only division.

For full directions and variations of this activity, see p. 74.

Activity

Things That Can't Be Shared Exactly

Hand out Student Sheet 11, Other Things to Share, one copy to each student. Students will be not be completing this page in a single activity, but will work on one problem at a time over Sessions 1 and 2, interrupting their work for a brainstorming activity and some money-sharing problems.

Read problem 1. Think about how you would share the brownies and the balloons. Use a drawing or picture to show your solutions, and write about your answers.

Students work individually on sharing 9 brownies and 9 balloons among 4 people. When they have finished, they briefly share their solutions with the class.

How large a share of brownies does each person get?

How did you deal with the extra balloon? Is it possible to cut a balloon in pieces and still have a balloon?

Activity

Can We Split It?

For this discussion activity, students set aside Student Sheet 11 while they explore things that can and can't be shared exactly. As a whole group, they brainstorm a list of things—like brownies—that we can split to share equally, and another list of things—like balloons—that we cannot split. Write headings on chart paper or on the board, and jot down ideas as students suggest them.

What are some things we can split? What can't we split? How could we share some of the things that we can't split?

❖ **Tip for the Linguistically Diverse Classroom** To ensure that everyone can understand the list, make simple rebus drawings next to or above each item you write down. Encourage students with very limited English proficiency to offer their own ideas by pointing to actual objects in the room or by making quick drawings themselves.

Classrooms that did this activity have included the following items in their lists. (You or your students might not agree with all of them!)

Activity

Sharing Dollars

Students explore something that sometimes can and sometimes can't be split equally: money. They work at first without a calculator, although they can use play money if available—dollars, quarters, and dimes. Pose a series of problems orally, or write them on the board:

Imagine that you and three friends share one dollar equally. How much money should each of you get? Write the answer using a dollar sign.

How much would you get if you shared two dollars among eight people?

Try a harder problem. How much would each person get if you share five dollars among four people?

Observe what students write down for each problem. If they cannot figure out how to write amounts of money with the dollar sign, show them some different ways to write money, such as $1.00, $1.25, $0.25 (or $ .25), and 25¢. Explain that we don't use a decimal point and ¢ together. To write 25 cents, we write either 25¢ or $ .25, which means one *quarter*, or one-fourth of a dollar. And $ .75 means three-fourths of a dollar—three quarters.

When students have figured out the answers to the three problems above and have written them in dollar notation, pass out calculators. Students should talk together and experiment with their calculators until they can do the same problems and get the answers they expect.

Note: You may need to remind students about the order of the numbers in division problems. These three problems are (without the dollar sign, which is not a calculator key) 1.00 ÷ 4, 2.00 ÷ 8, and 5.00 ÷ 4.

After students have had a chance to work with the calculator on these same problems, collect answers and observations in a brief discussion. They should have confirmed their answers as 0.25, 0.25, and 1.25.

For these problems, how did you write each person's share of the money? What did you notice about the calculator solutions?

Students might notice the 0.25 in two of the answers and talk about what they think that means. See the **Dialogue Box,** Sharing Money (p. 67), for an example of students talking together as they do these problems.

How the Calculator Writes One-Half Now present another problem to try on the calculator:

Six people share three dollars. How much money does each person get? Do this problem mentally and then do it on the calculator.

In working this problem, students will discover that the calculator shows one-half as 0.5 instead of as 0.50. Because they will have figured mentally that each person in this problem gets $ .50, students may wonder why the decimal answer has only one digit to the right of the decimal point. They may also be confused that the decimal for the smaller amount, one-fourth, has two digits, while the decimal for the larger fraction, one-half, has only one. If students don't mention this fact, raise the issue for discussion:

The smaller fraction, one-fourth, has two numbers (digits) after the decimal point—0.25. But one-half only has one number (digit)— 0.5. Why do you think that is?

After accepting students' ideas, show that 0.5 and 0.50 both mean one-half, because 0.5 is 5 out of 10 and 0.50 is 50 out of 100. Write $0.50, 0.5, 0.50, and 50¢ on the board.

Just as $0.50 means one-half of a dollar, 0.50 (or .50) is a way of writing one-half. The calculator doesn't put a zero at the end of decimals—it shows one-half in the simplest way, with just 0.5.

At this point, students need only recognize that 0.25 or .25 means one-fourth, or 25 out of 100, and that 0.5 and 0.50 both mean one-half. Don't worry if students don't yet understand how tenths and hundredths can be the same.

Completing the Student Sheet Students now return to Student Sheet 11, Other Things to Share, to complete problems 2 and 3. For problem 2, they may use either drawings or play money to determine how much money each person gets if 9 dollars is shared among 4 people, before checking their answers on the calculator.

The last problem asks students to circle all four of their answers on the sheet and think about how they are different and how they are alike. You might have students briefly share their thinking. Encourage them to explain how each different way of writing the answer goes with the particular problem it answers.

❖ **Tip for the Linguistically Diverse Classroom** For students with limited English proficiency, you might deal with problem 3 orally, asking them to respond by giving one-word answers to questions that address the similarities and differences of their answers to problems 1 and 2. For example: "In problem 1, did the people sharing brownies receive the same share as the people sharing balloons? Which two answers look the most alike? Which one answer is different from all the others? Which problems have an answer in decimal form?"

Activity

Fractions and Decimals

On chart paper or the board, begin a list of fraction and decimal equivalents. You may want to group the problems by denominator. Students can add to the lists as they explore on their own and when they do the Ten-Minute Math activity (Broken Calculator) over the next few days.

Fractions and Decimals

$\frac{1}{2} = 0.5$ $\frac{1}{4} = 0.25$ $\frac{1}{8} = 0.125$

$\frac{2}{4} = 0.5$ $\frac{2}{8} = 0.25$

$\frac{3}{4} = 0.75$ $\frac{3}{8} = 0.375$

Activity

Assessment

Sharing With and Without a Calculator

Student Sheet 12, Sharing With and Without a Calculator, presents some sharing problems similar to those students solved in Investigations 1 and 2. After students do each problem mentally or with drawings, they check it on the calculator. They also write a problem that has an answer of one-half and another with an answer of one-fourth.

Students might use pattern blocks to make up a problem with a share of one-half, or they might draw two or three brownies cut in half (and, for the later problem, cut in quarters). For each problem they make up, they should specify how many people can have a share.

❖ **Tip for the Linguistically Diverse Classroom** Students can record their made-up problems using quick sketches.

For things to consider when evaluating students' work, see the **Teacher Note**, Assessment: Sharing With and Without a Calculator (p. 66).

Sessions 1 and 2 Follow-Up

If students have calculators at home, they can revisit some of the problems they did in Investigation 1, such as how 5 brownies can be shared among 4 people. Have them find their completed copies of Student Sheet 3, Sharing Several Brownies, in their folders. They can now try the same problems on the calculator, writing the calculator answers on the same student sheet. Students then write about why the answers make sense or what they still wonder about.

Teacher Note

Assessment: Sharing With and Without a Calculator

Student Sheet 12, Sharing With and Without a Calculator, helps you assess the students' understanding of creating fractional shares and then representing fractions as decimals, using the calculator. Consider the following as you evaluate students' progress:

- Can students do the fraction sharing problems and write the answers in a way that is recognizable? You should expect that all students can do this after completing the *Fair Shares* unit.
- Can students divide using the calculator, getting the numbers in the right order? This task is new in these sessions. Some students may still find it confusing. The typical mistake is to turn the numbers around, so that two people sharing three brownies becomes $2 \div 3$. The student gets a confusing answer, and may say that a calculator can't do this kind of problem. Remind students that in a fraction, the line means to divide the numerator (top number) by the denominator (bottom number).
- Can students make up a problem with an answer of one-half and another with an answer of one-fourth? These are like the backward problems students did with pattern blocks in Investigation 2. Students may make up a problem based on pattern-block cookies, in which each person gets a share of one-half, but pattern blocks will not work for the share of one-fourth. Instead, they might think in terms of brownies, their Fraction Cards, or other simple drawings.
- Can students explain the meaning of both fraction and decimal answers? Some students will be able to explain the fraction and decimal answers separately but not compare them. They can explain 1 1/2 as one and a half brownies. They may explain 1.5 as one and a half dollars if you "add on a zero." They know that 0.25 is twenty-five cents or even a quarter of a dollar, but they do not yet think of it as equal to one-fourth in all circumstances.
- Can they explain why the fraction and decimal answers look different but mean the same thing? Many students will struggle with this concept. It is not important that students clearly explain why these answers look different in the third grade; their understanding of fractions and decimals will develop further each year.

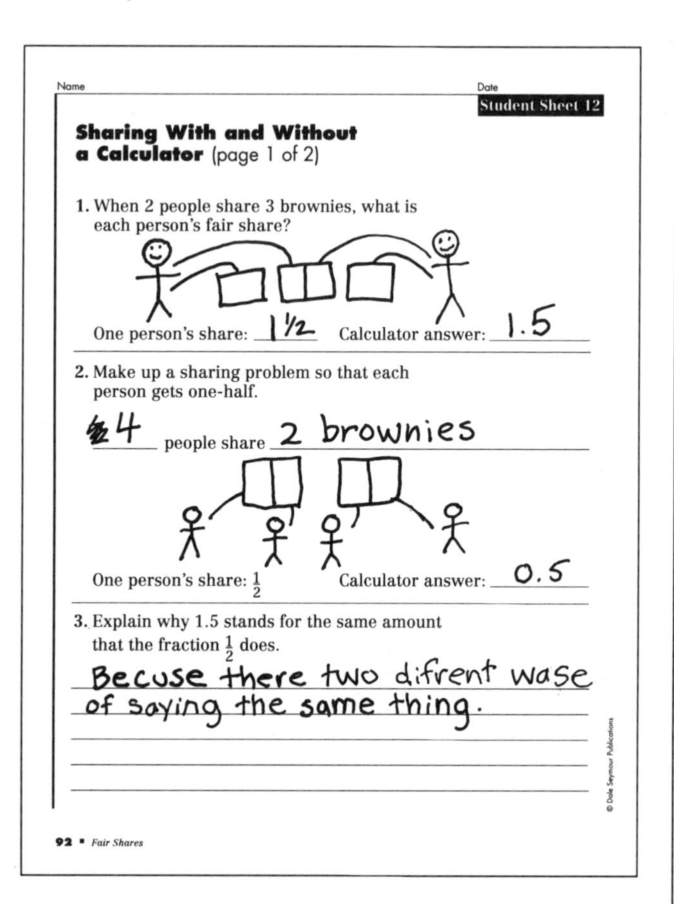

Name Date

Student Sheet 12

Sharing With and Without a Calculator (page 1 of 2)

1. When 2 people share 3 brownies, what is each person's fair share?

One person's share: 1 1/2 Calculator answer: 1.5

2. Make up a sharing problem so that each person gets one-half.

4 people share 2 brownies

One person's share: $\frac{1}{2}$ Calculator answer: 0.5

3. Explain why 1.5 stands for the same amount that the fraction $\frac{1}{2}$ does.

Becuse there two difrent wase of saying the same thing.

92 ▪ *Fair Shares*

© Dale Seymour Publications

Sharing Money

Here's how one group of students experimented with the calculator during the Sharing Dollars activity (p. 62).

Sean: One dollar divided among four people... [*He enters 1.00 ÷ 4 into the calculator.*] Twenty-five cents, and it's zero point twenty-five cents. What is two dollars shared among eight people?

Yvonne: Twenty-five cents. It's the same. Since eight is twice four, I doubled both things, so it has to be the same.

Rashad: Oh yeah!

Yvonne: Let me do it on the calculator. [*She presses the numbers.*] Zero point two five.

Rashad: Let's make one up. Three dollars shared among four people.

Yvonne: Fifty cents each, and a dollar is left over, and I give each one a quarter.

Su-Mei: So that's seventy-five cents? I think so too, but when I check it on the calculator, it keeps coming out to two point twenty-five (2.25).

Rashad: [*does the problem on the calculator, 3.00 ÷ 4*] Yup. It is seventy-five cents!

Su-Mei: Yes, but look. I was doing seventy-five *plus* seventy-five [*entering 75 + 75 = again on the calculator*]. It's one dollar and fifty cents, plus seventy-five ... is two dollars and twenty-five cents. So what's happening?

Tell me how many seventy-fives are in two hundred twenty-five.

Su-Mei: [*keeping track on her fingers*] Seventy-five, one hundred fifty, two hundred twenty-five. Oh, only three! That's what happened!

Session 3

Sharing Many Things

Materials

- Small items to share, such as paper clips, pennies, or peanuts (8–10 per student)
- Calculators

What Happens

Students determine how to share a large number of items equally with their group. They determine what fraction of the items each student should get, and then figure out how many of the items the fraction represents. Their work focuses on:

- identifying the fraction that represents equal shares
- using fractions for division

Ten-Minute Math: Broken Calculator Several times during the next few days, continue the Broken Calculator activity, with students trying to get decimal numbers on the calculator display using division and no decimal point.

Collect division problems and answers from students. Group problems that have the same answers, starting with 0.5, 0.25, and 0.75 (if any students have invented problems that give 0.75, or are ready to try it). Ask students to look for what is alike in all the problems that give a certain answer. (For 0.5, a number is always divided by twice itself; for 0.25, a number is always divided by four times itself.)

When students have identified some patterns, they can do this activity with partners. The first partner starts the math sentence, and the second person finishes it to come up with 0.5 (or any other decimal they agree to in advance) as the answer. Illustrate this activity with all students entering the same number, perhaps 10, and then pressing ÷ and whatever number they think will get 0.5. For example, ask:

Ten divided by what number will make point five (five-tenths)?

Partners working together might investigate what other decimals they get by dividing 10 by other multiples of 10, or dividing 1 by small numbers. Once they get a new decimal, they can work together to find other numbers they can divide to make that same decimal. Collect problems from students, grouping together those that make the same decimal.

Activity

How Many for Each Person?

Give each group a pile of pennies, peanuts, or other small items to share. Choose a number of items that will *not* divide evenly among the group. To save preparation time, students can count out the number of items you suggest and return the extras to you.

Students first decide what fraction each person in their group will get if they divide the pile into fair shares. The students need to agree on the appropriate fractional amount for each share, then decide how many items that would be. (If you use peanuts in shells, tell students not to open the peanuts to split them for sharing.)

What fraction of the [peanuts] do you think each of you should get?

Figure out in your minds how many [peanuts] each person in the group will get if you share them equally. Then use the [peanuts] to test your solution.

If some students suggest sharing out their items into groups right away, acknowledge that they could get the exact answer that way, but point out that you first want them to try to figure it out without actually sharing.

Then students can work out this problem in their groups, using the items you have provided.

Other Sharing Problems After completing this problem in small groups, students can do other problems of this kind individually, working out their answers in writing and with drawings. For example:

> Find $\frac{1}{3}$ of 40 beads, to show how many beads each of 3 people could have.
>
> Find $\frac{1}{10}$ of 85 paper clips, so that 10 people can share the paper clips equally.

When most students have finished writing, provide some time for students to share their approaches. If students do not mention the similarity between fractions and division, you can use their solutions as the basis for asking them about it.

What are some different ways we can write the same problem?

If you have previously used the third grade Multiplication and Division unit, *Things That Come in Groups*, students may recognize any of these ways of representing the problem:

20 has how many 4's? $\frac{1}{4}$ of 20

$20 \div 4$ $\frac{20}{4}$ $4\overline{)20}$

Activity

Choosing Student Work to Save

As the unit ends, you may want to use one of the following options for creating a record of students' work on this unit.

- Students look back through their folders or notebooks and write about what they learned in this unit, what they remember most, what was hard or easy for them. You might have students do this work during their writing time.
- Students select one or two pieces of their best work. You also choose one or two pieces of their work to be saved in a portfolio for the year. You might include students' written solutions to the assessments in this unit—Letter to a Second Grader (p. 44) and Sharing With and Without a Calculator (p. 65). Students can create a separate page with brief comments describing each piece of work.
- You may want to send a selection of work home for parents to see. Students write a cover letter, describing their work in this unit. This work should be returned if you are keeping a year-long portfolio of mathematics work for each student.

Session 3 Follow-Up

■ **Baking a Snack to Share** Bake a batch of brownies or corn bread in a large rectangular pan for students to share. Have each pair of students draw the pan, showing where they would make cuts to make shares for you and all the students in the class. Choose one plan that makes sense to you, and invite that student pair to cut the snack. Advise the partners to place their cuts carefully, because they will be the last to choose pieces for themselves.

■ **More Broken Calculator** After finishing this Fractions unit, continue the Broken Calculator activity a few times in the next few weeks. Ask students to explore what decimals they can get by using only the division key and single-digit numbers.

Use the calculator to make different decimal numbers. Use only the division key and numbers smaller than 10. (You can use bigger numbers later.) On paper, keep track of the numbers you press and the answers you get.

Try to make up problems that come out fairly evenly, and not the ones like "point three three three three three" (0.33333). For example, if I press 3, ÷, 6, =, what do I get?

After students have answered, write this on the board as both a division problem and as a fraction-decimal equivalent:

$3 \div 6 = 0.5$ $\frac{3}{6} = 0.5$

Keep a classroom list of the numbers the students divide and the decimals they get. Write their problems both as division problems and as fractions. Group together problems that have the same decimal answer. Ask students how you might further organize the list to see patterns (for example, by grouping their answers by denominator).

Guess My Number

Basic Activity

You choose a number for students to guess, and start by giving clues about the characteristics of the number. For example: It is less than 50. It is a multiple of 7. One of its digits is 2 more than the other digit.

Students work in pairs to try to identify the number. Record students' suggested solutions on the board and invite them to challenge any solutions they don't agree with. If more than one solution fits the clues, encourage students to ask more questions to narrow the field. They might ask, for example: Is the number less than 40? Is the number a multiple of 5?

Guess My Number involves students in logical reasoning as they apply the clues to choose numbers that fit and to eliminate those that don't. Students also investigate aspects of number theory as they learn to recognize and describe the characteristics of numbers and relationships among numbers. Students' work focuses on:

- systematically eliminating possibilities
- using evidence
- formulating questions to logically eliminate possible solutions
- recognizing relationships among numbers, such as which are multiples or factors of each other
- learning to use mathematical terms that describe numbers

Materials

- 100 chart (or 300 chart, when a larger range of numbers is being considered)
- Scraps of paper or numeral cards for showing solutions (optional)
- Calculators (for variation)

Procedure

Step 1. Choose a number. You may want to write it down so that you don't forget what you picked!

Step 2. Give students clues. Sometimes, you might choose clues so that only one solution is possible. Other times, you might choose clues so that several solutions are possible. Use clues that describe number characteristics and relationships, such as factors, multiples, the number of digits, and odd and even.

Step 3. Students work in pairs to find numbers that fit the clues. A 100 chart (or 300 chart for larger numbers) and scraps of paper or numeral cards are useful for recording numbers they think might fit. Give students just one or two minutes to find numbers they think might work.

Step 4. Record all suggested solutions. To get responses from every student, you may want to ask students to record their solutions on scraps of paper and hold them up on a given signal. Some teachers provide numeral cards that students can hold up to show their solution (for example, they might hold up a 2 and a 1 together to show 21). List on the board all solutions that students propose. Students look over all the proposed solutions and challenge any they think don't fit all the clues. They should give the reasons for their challenges.

Step 5. Invite students to ask further questions. If more than one solution fits all the clues, let students ask yes-or-no questions to try to eliminate some of the possibilities, until only one solution remains. You can erase numbers as students' questions eliminate them (be sure to ask students to tell you which numbers you should erase). Encourage students to ask questions that might eliminate more than one of the proposed solutions.

Variations

New Number Characteristics During the year, vary this game to include mathematical terms that describe numbers or relationships among numbers that have come up in mathematics class. For example, include factors, multiples,

Continued on next page

doubling (tripling, halving), square numbers, prime numbers, odd and even numbers, less than and more than concepts, as well as the number of digits in a number.

Large Numbers Begin with numbers under 100, but gradually expand the range of numbers that you include in your clues to larger numbers with which your students have been working. For example:

- It is a multiple of 50.
- It has 3 digits.
- Two of its digits are the same.
- It is not a multiple of 100.

Guess My Fraction Pick a fraction. Tell students whether it is smaller than one-half, between one-half and one, between one and two, or bound by any other familiar numbers. You might use clues like these:

- It is a multiple of one-fourth (for example, one-half, three-fourths, one whole, one and one-fourth).
- The numerator is 2 (for example, two-thirds, two-fifths).
- You can make it with pattern blocks (for example, two-thirds, five-sixths).

Don't Share Solutions Until the End As students become more practiced in formulating questions to eliminate possible solutions, you may want to skip step 4. That is, student pairs find all solutions they think are possible, but these are not shared and posted. Rather, in a whole-class discussion, students ask yes-or-no questions, but privately eliminate numbers on their own list of solutions. When students have no more questions, they volunteer their solutions and explain why they think their answer is correct.

Calculator Guess My Number Present clues that provide opportunities for computation using a calculator. For example:

- It is larger than 35×20.
- It is smaller than $1800 \div 2$.
- One of its factors is 25.
- None of its digits is 7.

Related Homework Options

Guess My Number Homework Prepare a sheet with one or two Guess My Number problems for students to work on at home. As part of their work, students should write whether they think only one number fits the clues or whether several numbers fit. If only one solution exists, how do they know it is the only number that fits the clues? If more than one solution is possible, do they think they have them all? How do they know?

Students' Secret Numbers Each student chooses a number and develops clues to present to the rest of the class.You'll probably want to have students submit their numbers and clues to you for review in advance. If the clues are too broad (for example, 50 solutions are possible) or don't work, ask the students to revise their clues. Once you approve the clues, students are in charge of presenting them, running the discussion, and answering all questions about their number during a Ten-Minute Math session.

Broken Calculator

Basic Activity

Students work to get an answer on their calculator display while pretending that some of the keys are missing. The missing keys can be operations, numbers, or both. After students find one solution, they find others by making a small change in the first one. In this way, the solutions form a pattern.

Broken Calculator helps students develop flexibility in solving problems. They pull numbers apart and put them back together in a variety of ways as they look for expressions to substitute for given numbers. Students focus on:

- finding alternative paths to an answer when a familiar one isn't available
- finding many ways to get one answer
- writing related problems

Materials

Calculators: 1 per student

Procedure

Step 1. Pose the problem. For example, "I want to make 35 using my calculator, but the 3 key and the 5 key are broken. How can I use my calculator to do this task?"

Step 2. Students solve the problem by themselves. They record their solution in some way that another student can understand. Students in small groups check each others' solutions on their calculators.

Step 3. List some of the students' solutions on the board. Here are some possible solutions to making 35 without the 3 and 5 keys:

$76 - 41$ $29 + 6$ $4 \times 9 - 1$

Step 4. Students choose one solution and extend it, making a series of related solutions that follows a pattern. For example:

76 − 41	29 + 6	2 × 18 − 1
77 − 42	28 + 7	4 × 9 − 1
79 − 44	27 + 8	6 × 6 − 1
81 − 46	26 + 9	
82 − 47	24 + 11	

Students check each others' solutions and find another solution that follows the same pattern.

Variations

Restricting Number Keys

- Students make numbers without using the digits in those numbers, for example:

  Make 1000 without using a 1 or a 0.

  998 + 2

  997 + 3

  996 + 4

- Students make decimals without using the decimal point. Start with the simplest ones (0.1, 0.5, 0.25, 0.75, or 1.5) only after students have some experience relating them to fractions and division. You might start by providing a solution or two and challenge them to find some more: "I can make 0.5 on my calculator by using the keys 1 ÷ 2. Why do you think that works? Can you find another way to make 0.5?"

  Some solutions for making 0.5 are as follows:

  2 ÷ 4 3 ÷ 6 4 ÷ 8 5 ÷ 10

  100 ÷ 200 1000 ÷ 2000.

Restricting Operation Keys

- Students make a number using only addition. If you suggest a large number, students can make use of landmark numbers. For example:

  Make 2754.

2000 + 700 + 54	2750 + 4	2749 + 5
2000 + 600 + 154	2751 + 3	2748 + 6

- Students make a number using only subtraction. The +, ×, ÷ keys are broken. Patterns of solutions for making 8 might look like these:

20 − 12	1008 − 1000
19 − 11	908 − 900
18 − 10	808 − 800
17 − 9	708 − 700

- Students make a number using only multiplication and division. The + and – keys are broken. Pick numbers that have many factors. Answers for making 24 might be:

1×24	$24 \div 1$	$24 \times 1 \div 1$
2×12	$48 \div 2$	$24 \times 2 \div 2$
3×8	$72 \div 3$	$24 \times 3 \div 3$
4×6		$24 \times 4 \div 4$

(One student filled a page with the third series so he could say he'd gotten the most answers.)

Restricting Both Operations and Digits

Make the missing operations problems more challenging by also not allowing students to use any of the digits in the final number. For example:

> Make 654 using only addition and subtraction, and without using the digits 6, 5, or 4.

Related Homework Option

Pose one or two Broken Calculator problems only. Challenge students to solve the problems in more than one way, and to make their different solutions follow a pattern. They should write down their solutions so that another student can read them and know what to do on the calculator.

If students do not have calculators at home, give them time to try out their solutions the next day in school.

The following activities will help ensure that this unit is comprehensible to students who are acquiring English as a second language. The suggested approach is based on *The Natural Approach: Language Acquisition in the Classroom* by Stephen D. Krashen and Tracy D. Terrell (Alemany Press, 1983). The intent is for second-language learners to acquire new vocabulary in an active, meaningful context.

Note that *acquiring* a word is different from *learning* a word. Depending on their level of proficiency, students may be able to comprehend a word upon hearing it during an investigation, without being able to say it. Other students may be able to use the word orally, but not read or write it. The goal is to help students naturally acquire targeted vocabulary at their present level of proficiency.

We suggest using these activities just before the related investigations. The activities can also be led by English-proficient students.

Investigations 1–3

fair, share, cookie

1. Show a bag of cookies. As you take a cookie out, ask who (pointing to students) would like to *share* it with you.
2. After students enthusiastically raise their hands, break off a very small crumb from the cookie and hand it to one student. Point out that you did not offer a *fair* share.
3. This time, pull out of the bag one cookie for every two students. Ask the students to show you how they might share their cookies so that each person has the same amount. Emphasize that each partner must have a *fair* share. After students have broken their cookies, ask them to decide if everyone has shared fairly.

   Do Su-Mei and Annie both have fair shares of the cookie? Do Ryan and Khanh both have fair shares?

whole, piece

1. Hold up one cookie and identify it as a *whole* cookie. Then break the cookie, and identify the parts as *pieces* of cookie. Do the same actions with a paper brownie, cutting it into pieces.
2. Challenge students to demonstrate comprehension of the words *whole* and *piece* by showing other examples of both and asking questions with one-word answers:

   Is this a whole banana or a piece of banana?

   Is this a whole carrot or a piece of carrot?

   Is this a whole puzzle or a piece of a puzzle?

brownie, smallest, largest

1. Show a plate of brownies, or a picture of brownies from an advertisement or box of baking mix. Relate them to the rectangles of paper students will be calling "brownies" in the activities.
2. Prepare about 6 paper rectangles of different sizes and identify them as "brownies." Explain that you want to arrange the paper brownies from smallest to largest. Use your fingers to convey "small," and then identify the smallest paper brownie. Ask students to identify which one would come next, in a ranking from smallest to largest.

   This is the smallest brownie. Which brownie is next smallest?

   Continue until all the brownies have been ranked.

Blackline Masters

________________, 19 ____

Dear Family,

Our class is starting a unit on fractions, called *Fair Shares*. Your child will be using fractions to make "fair shares" of things like paper "brownies" and pattern-block "cookies."

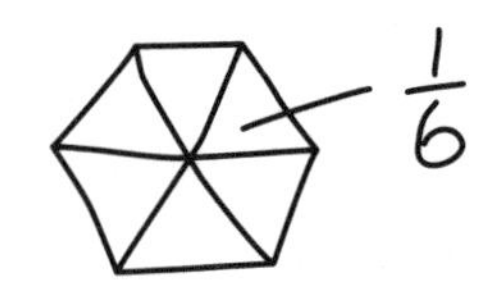

By working with these materials, your child will learn about how "wholes" come apart into fractions, and fractions fit together as wholes. We'll talk about which of two shares is larger, and which fractions are really the same (for example, $1/3$ is the same as $2/6$). Here are some ways that you can help at home:

- When your family really is sharing food, talk about "fair shares" and help your child name fractions. If you want to split the pizza among five people, how can you split it fairly? How much does each person get?
- Cooking is another great way to learn about fractions. How can we measure $3/4$ cup? Look together at how the fractions appear on a measuring cup. Doubling recipes, or cutting them in half, can help your child understand how to make new numbers with fractions.
- Throughout the unit, look over your child's homework. Ask about the fraction work the class is doing, and encourage your child to explain some of the problems to you. Early on, you'll make a set of Fraction Cards together. Later, your child will learn a game you can play with these cards.

You may be surprised that your child won't be learning step-by-step procedures for working with fractions. Many adults remember the "invert and multiply" rule they learned for dividing fractions, but few can explain how and why this works. Your children will learn the hows and whys—and, hopefully, will become much more comfortable using fractions.

Sincerely,

Name Date

Student Sheet 1

Sharing One Brownie (page 1 of 2)

Cut up large brownie-rectangles and glue the pieces below.
Show how you would make fair shares.

1. 2 people share a brownie. Each person gets 

2. 4 people share a brownie. Each person gets 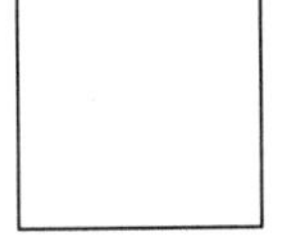

3. 8 people share a brownie. Each person gets 

Sharing One Brownie (page 2 of 2)

4. 3 people share a brownie. Each person gets ☐

5. 6 people share a brownie. Each person gets ☐

Name _______________________________ Date

Student Sheet 2

Different-Shaped Pieces

Here is a picture of a brownie cut into four pieces:

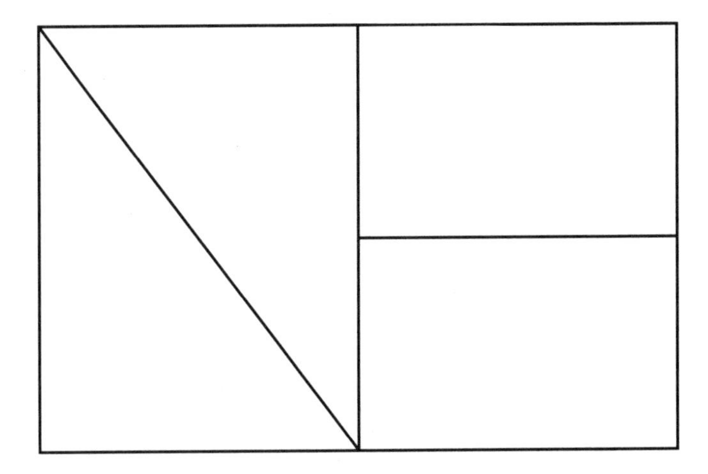

Some people think these are not fair shares.
Write what you believe.

I believe that the rectangle pieces and the triangle pieces ______________ the same size because ...
are / are not

____

____

____

____

____

____

____

____

____

____

Name Date

Student Sheet 3

Sharing Several Brownies

________ brownies shared by ________ people
number of brownies number of people

One person's share is ________.

Name Date

Naming Fraction Shares

1. In these brownie pictures, one person's share is the shaded part.

a. How much brownie is this share?

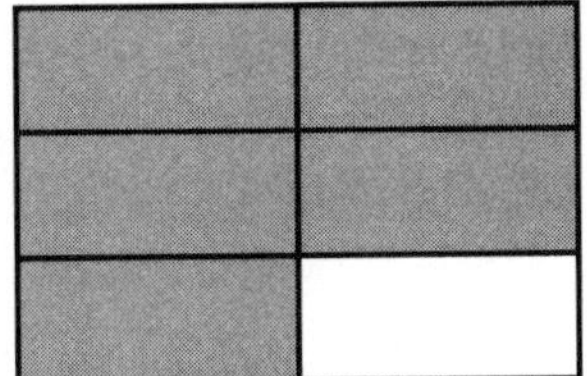

b. How much brownie is this share?

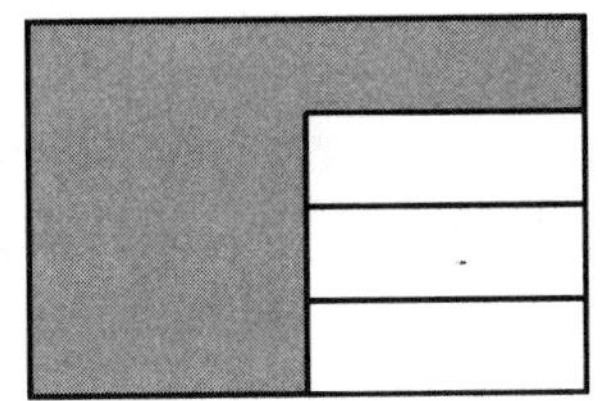

2. Three students did this problem:

They got three different answers for one person's share.

All of their answers are correct.

How can that be?

Make drawings to show that at least two of these answers are the same.

3 people sharing 5 brownies	
Anne:	$1\frac{2}{3}$
Rachel:	1 and $\frac{1}{2}$ and $\frac{1}{6}$
Carl:	$\frac{5}{3}$

Name

Date

Student Sheet 5

Hexagon Cookies

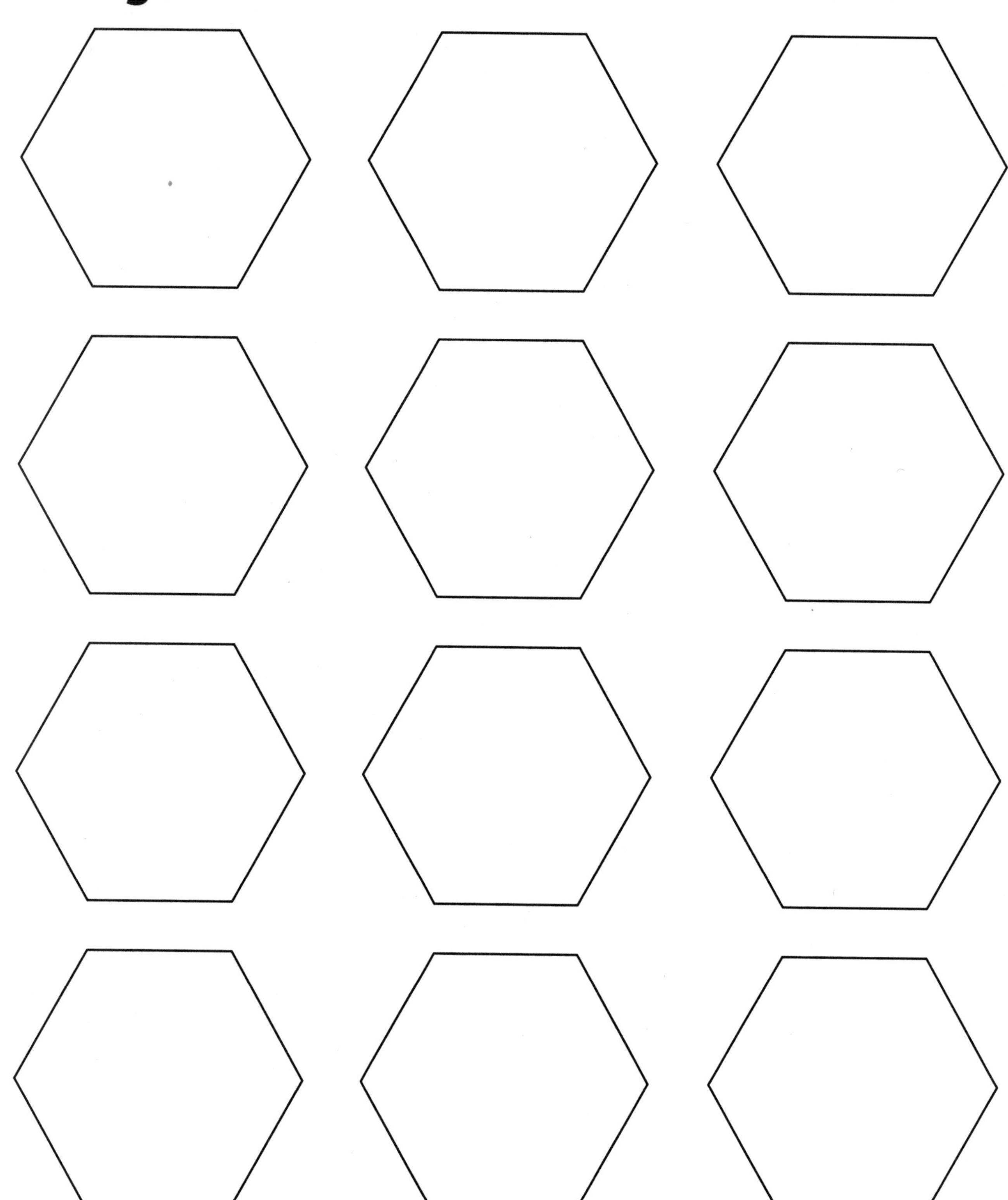

Use pattern blocks. Show all the ways to make 1 whole cookie. Have you found them all? Are any of your designs the same?

Name Date

Many Ways to Make a Share

Think of sharing brownies or hexagon cookies.
Write all the fractions you know that work.

Ways to make 1 whole	Ways to make $\frac{1}{2}$
Ways to make $\frac{1}{3}$	Ways to make $\frac{2}{3}$
Ways to make $\frac{1}{4}$	Ways to make $\frac{3}{4}$
Challenge: Ways to make $\frac{5}{6}$	**Challenge:** Ways to make $\frac{5}{8}$

Name Date

Student Sheet 7

Who Gets the Larger Share? (page 1 of 2)

1. Group A: 3 people share 5 brownies.
   Group B: 2 people share 5 brownies.

   Who gets the larger share? ______________________
   Tell how you decided. Use words or drawings or both.

---

2. Group C: 6 people share 4 brownies.
   Group D: 3 people share 2 brownies.

   Who gets the larger share? ______________________
   Tell how you decided. Use words or drawings or both.

Who Gets the Larger Share? (page 2 of 2)

3. Group E: 3 people share 4 cookies.
Group F: 2 people share 3 cookies.

Who gets the larger share? ____________________
Tell how you decided. Use words or drawings or both.

---

4. Group G: 3 people share 8 cookies.
Group H: 2 people share 7 cookies.

Who gets the larger share? ____________________
Tell how you decided. Use words or drawings or both.

Name Date

Student Sheet 8

Letter to a Second Grader

Some people think 1 and $\frac{1}{4}$ is a larger share than 1 and $\frac{1}{3}$.

Some people think 1 and $\frac{1}{3}$ is a larger share than 1 and $\frac{1}{4}$.

Which share do you think is bigger?
Write a letter to a second grader. Tell why you are right.
Use drawings to explain your thinking.
Remember, a second grader must understand what you write.

Name Date

How to Play the Fraction Card Game

Materials: 2 sets of Fraction Cards

Players: 2 (or two teams of 2)

How to Play

1. This game requires some space, so try to play on the floor or find a table that is clear. Each player starts with a full set of Fraction Cards that are mixed up. Turn the cards so the fraction labels are face down.

2. Both players turn up one card at a time from the top of their sets, showing the fraction label.

   The player who turns up the bigger Fraction Card takes both cards (like in the card game War). If both cards are the same size, both players turn up another card. The player with the bigger fraction takes all four cards.

3. Each time you win cards, try to make a whole with them. You may use any cards you have already won.

   For example, you could make some wholes out of halves, thirds, and sixths. You could make other wholes out of halves, fourths, and eighths.

   Other combinations will also work together, such as 2 fourths and 3 sixths. Leave your wholes out on the table.

4. Continue taking turns and making wholes until you have used up all the cards. The person with the most wholes at the end wins the game.

Name

Date

Student Sheet 10

How Many Altogether?

1. Isaac had a birthday party. He baked large cookies for himself and five friends. After the six people at the party shared all the cookies evenly, each person had 1 and $\frac{1}{3}$ cookies.

   How many cookies did Isaac bake altogether? Show your work here or on the back.

2. Four friends shared some brownies evenly. Each person got a whole brownie and one-fourth of another brownie.

   How many brownies did the friends start with? Show your work.

3. Three friends shared some cookies. They each got two cookies and two-thirds of a cookie.

   How many cookies did they have altogether? Show your work.

Challenge Problem

A group of friends had some cookies that they shared evenly. Each person got one and a half cookies.

How many cookies do you think they might have started with? How many people might have been in the group?

Number of people	Number of cookies

Name Date

Other Things to Share

1. How would you share each of the following? Write about your thinking or use a drawing to show your solution.

9 brownies shared among 4 people

9 balloons shared among 4 people

---

2. How much money does each person get? Compare your answer to the answer you get using a calculator.

9 dollars shared among 4 people

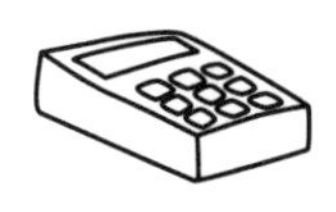

$9 \div 4$ on a calculator

---

3. Put a circle around each of your four answers above. They are all answers to 4 people sharing 9 things.

How are they different from each other? How are they alike?

Name

Date

Student Sheet 12

Sharing With and Without a Calculator (page 1 of 2)

1. When 2 people share 3 brownies, what is each person's fair share?

One person's share: ________ Calculator answer: ____________

2. Make up a sharing problem so that each person gets one-half.

________ people share ____

One person's share: $\frac{1}{2}$ Calculator answer: ____________

3. Explain why 1.5 stands for the same amount that the fraction $\frac{1}{2}$ does.

____

____

____

____

____

Sharing With and Without a Calculator (page 2 of 2)

4. When 4 people share 5 brownies, what is each person's fair share?

One person's share: ________ Calculator answer: ____________

5. Make up a sharing problem so that each person gets one-fourth.

______ people share ____________________________

One person's share: $\frac{1}{4}$ Calculator answer: ____________

6. What fraction is 0.25 equal to?

How would you explain to a friend why the two amounts are equal?

____

____

____

____

____

Large Brownies

Small Brownies

How to Make Fraction Cards

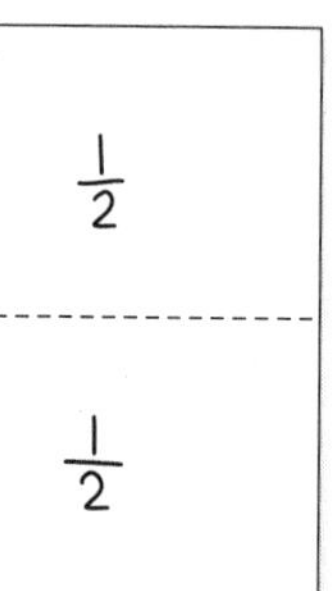

Materials: 5 sheets of paper (all the same color), a pen or crayon, scissors

What to do

Fold and label paper as shown. Write each fraction on one side only. Mark the fold lines. Cut on the lines.

Halves Fold the sheet in the middle.

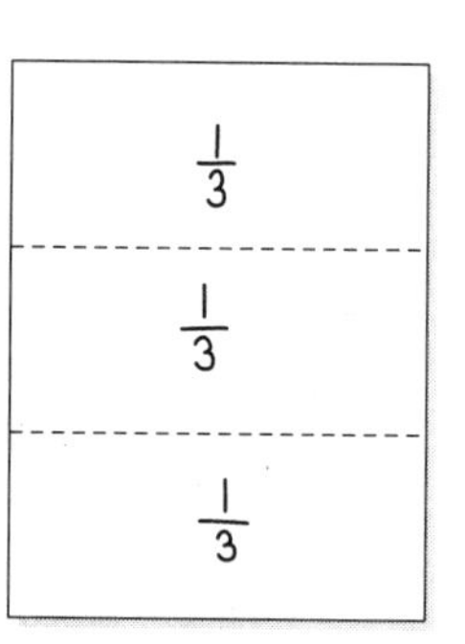

Thirds Fold the sheet in 3 equal pieces.

Fourths Fold the sheet in half one way, then in half the other way, to make 4 equal pieces.

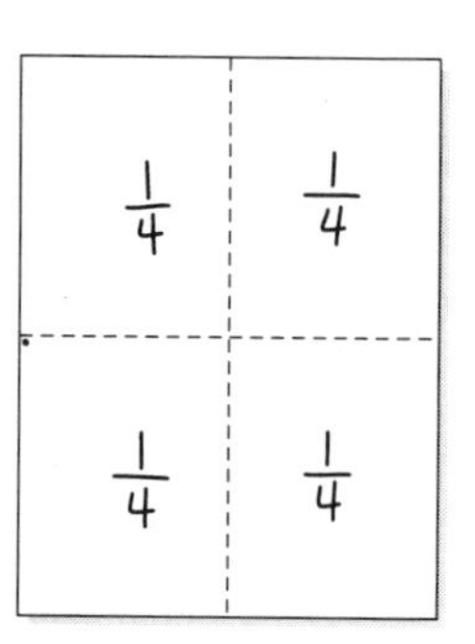

Sixths Fold the sheet the way you did to make thirds. Cut apart the thirds. Fold two of the thirds in half and cut to make long, skinny sixths. Cut the other third in half the other way to make chunky sixths.

Eighths Fold the sheet into fourths. Open the paper to see the folds. Fold each fourth in half.

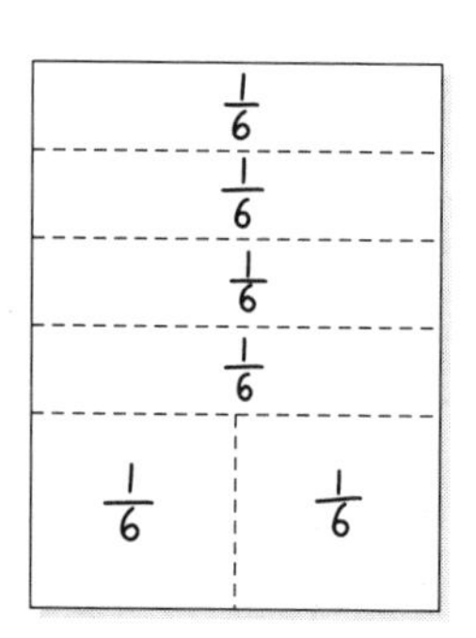

Store your Fraction Cards in a plastic bag or envelope. Save these directions with them. You may want to make another set for playing games.

Ideas to Try at Home

Turn over the cards to hide the labels. Order the cards from smallest to largest. Then turn them over. Look at the number pattern. What do you see?

Find different combinations to make one whole. Put the Fraction Cards on top of a whole piece of paper to keep track. Keep a list of your combinations. For example:

$\frac{1}{2} + \frac{1}{3} + \frac{1}{6} = 1$ $\qquad$ $\frac{1}{3} + \frac{1}{3} + \frac{1}{3} = 1$